WAR AS EXPERIENCE

This book is a major new contribution to our understanding of war and how it should be studied in the field of international relations (IR). It is divided into two sections. The first part surveys the state of war and war studies in IR, in security studies and in feminist IR, using an exemplary texts approach. The second part addresses a missing area of IR studies of war that feminism is well placed to fill in: the physical and emotional aspects of war.

The author demonstrates how war is experienced as a body-based politics and in so doing provides an innovative and challenging corrective to traditional theories of war. This will be essential reading for all those with an interest in gender, war, and international relations.

Christine Sylvester is Professor of Political Science and of Women's Studies at the University of Connecticut and is an affiliated professor of the School of Global Studies, The University of Gothenburg, Sweden.

Series: War, Politics and Experience
Series Editor: Christine Sylvester

Experiencing War
Edited by Christine Sylvester

The Political Psychology of War Rape
Studies from Bosnia and Herzegovina
Inger Skjelsbæk

Gender, Agency and War
The maternalized body in US foreign policy
Tina Managhan

War as Experience
Contributions from international relations and feminist analysis
Christine Sylvester

WAR AS EXPERIENCE

Contributions from international relations and feminist analysis

Christine Sylvester

Routledge
Taylor & Francis Group

LONDON AND NEW YORK

First published 2013
by Routledge
2 Park Square, Milton Park, Abingdon, Oxon OX14 4RN

Simultaneously published in the USA and Canada
by Routledge
711 Third Avenue, New York, NY 10017

Routledge is an imprint of the Taylor & Francis Group, an informa business

British Library Cataloguing in Publication Data
A catalogue record for this book is available from the British Library

Library of Congress Cataloging in Publication Data
A catalog record for this book has been requested

ISBN: 978-0-415-77598-4 (hbk)
ISBN: 978-0-415-77599-1 (pbk)
ISBN: 978-0-203-10094-3 (ebk)

Typeset in Bembo
by RefineCatch Limited, Bungay, Suffolk

Printed and bound in Great Britain by the MPG Books Group

To friends and colleagues at Gothenburg, Lund, and Malmö universities, Sweden.

CONTENTS

Acknowledgements   ix

Introduction: War questions for feminism and
International Relations   1

PART I
International Relations and feminists consider war   **15**

1   IR takes on war   17

2   Feminist (IR) takes on war   38

PART II
Rethinking elements and approaches to war   **63**

3   War as physical experience   65

4   War as emotional experience   87

5   Concluding, collaging, and looking ahead   111

Notes   127
References   131
Index   142

ACKNOWLEDGEMENTS

This book would have been difficult to write without the wonderful Kerstin Hesselgren Professorship I was awarded by the Research Council of Sweden; it enabled me to spend a stimulating academic year (2010-2011) writing away at the School of Global Studies (SGS), the University of Gothenburg. SGS is one of the more innovative academic homes I have been fortunate to experience at close range, and for me part of its charm lies in my not quite realizing what a rare interdisciplinary place it is. I would like to offer very special thanks to Maria Stern, Stina Sundling Wingfors, and Sungju Park-Kang, my PhD student who joined me there.

Remarkably, I was then invited to stay on in Sweden for 2011-2012 at the Political Science Department at Lund University, and spent one semester there before moving to a terrific surprise professorship at the University of Connecticut. Many thanks to Annica Kronsell and Tomas Bergstrom at Lund and to Mark Boyer, Nancy Naples, and Jeremy Teitelbaum at UConn; I am now back in my home state of Connecticut after eighteen years working in Australia, the Netherlands, Britain, and Sweden.

Three people graciously read all or part of this book manuscript, something I genuinely prize. Lene Hansen and Swati Parashar commented in detail on an entire draft – Lene taking part of her vacation period in Denmark to do so, and Swati squeezing the reading between the cricket matches she was avidly watching at the time in Australia. Megan MacKenzie commented on one chapter from her then perch in Wellington, New Zealand. Each of these worldly friends offered their special and inimitable takes on the work, for which I am most grateful. I also presented parts of the manuscript to colleagues at Kent University, the Peace Research Institute of Oslo, Lund University, Gothenburg University, the University of Connecticut, a London School of Economics Millennium Conference, and several International Studies Association conferences. I very much appreciate those invitations and the input of so many colleagues.

The incomparable Michael Weil repeatedly read my mind, repeatedly kept the computers running and the homefires burning, and kept shaking his head sympathetically while I tapped away. My old friend Karen Pugliesi watched Eliza dog in Arizona while I was in Lund (very heroic of her), new friends Dan and Nancy Lucente watched the house in Mystic whenever Michael and I were both away, and Erika Svedberg and family – Gion, Vera, Fabian, and Tilly dog – nurtured my soul in Sweden. To all of you, Skål!

INTRODUCTION

War questions for feminism and International Relations

War has decreased in frequency in our time. Based on counting armed conflicts between and within states, IR analysts can claim that we live in a less warlike era than, say, during the Cold War period of the twentieth century (Gleditsch, 2008; Gleditsch *et al.*, 2002; Newman, 2009; Mueller, 2004; Goldstein, 2011). Of course, that finding relies on certain assumptions, a key one often being that states are the main actors involved in war, the key participants. Such studies can also assume that casualties are counted properly (see Butler, 2004b; Melander *et al.*, 2007) and that war takes recognizable forms. Yet it is abundantly clear that war is trickier to recognize, count, and tally up than it was in earlier times or under earlier state-centric understandings of war.[1] There are so many participants in today's wars that it can be difficult to determine which one among them is "the" main actor whose presence determines that war is taking place; at the very least, there can be guerrilla forces, networked organizations, private firms, states, and mixed state and nonstate coalitions involved (Abramson and Williams, 2010; Leander, 2012; Stanger, 2009). Wars also always involve people, whose locations with respect to war, and experiences with it, are diverse and yet significant for them, their societies, international relations, short- and long-term outcomes of wars, and the prospect of future wars. And then there is the thorny problem of reckoning with casualties in war, which some IR accounts use as a measure of war intensity. The problem there, as Judith Butler (2004b) aptly reminds her readers, is that some war deaths are counted and grieved and many others are ignored – because deaths of everyday people in wars "over there" are collateral rather than important damage for the counters to tally.

This book argues that understanding people's experiences with/in war is essential for understanding war. Feminist IR war studies of recent vintage make that proposition a touchstone of their research. The rest of IR tends to operate at more abstract levels of analysis when studying war, focusing on states, organizations, laws, norms, discourse and the like. Yet even IR theories that seem most devoid of people can show eyes peeking through cracks in the analysis and gazing out from everyday locations in homes, villages, battle areas, fields, or streets. Sadly, those brief and often anecdotal moments showing that international relations is a place of people are overpowered in much of IR by a speedy return to abstract actors. Yet war cannot be fully apprehended unless it is studied up from people's physical, emotional, and social experiences, not only down from "high politics" places that sweep blood, tears, and laughter away, or assign those things to some other field.

It is not as easy to define war when the participants in it are conceptualized as multiple, scattered, and occupying "high" and "low" ground. Consider some examples. A state might take on a troublesome regime head (Saddam Hussein), a rump group that once ran a state (the Taliban), militant anti-state forces inside state borders, like the Tamil Tigers of Sri Lanka, or warlords of failed states (Somalia). Wars can also create and engage civilian identity groups that go at each other viciously, despite having lived together relatively comfortably as Yugoslavs or Rwandans up to that point. Some wars start with one goal and set of participants and end up in other social realms and spaces. The 2003 Iraq war, for instance, began as a high-tech operation of American military prowess – shock and awe – unleashed against Iraq to force regime change. That assault led to urban street battles in several locations across the country as religious and secular militants responded to the attacks and to each other. It then turned into a complex civil war to settle old and new scores, which led to a sustained counterinsurgency effort by a coalition of forces attempting to set the country on a new path. As well, the American-led war on Afghanistan ostensibly defeated the Taliban in the first round, but that group rose again in mountain regions and remote towns, and the war spilled into neighboring Pakistan via a number of "discreet military operations" (Zenko, 2010), including the one that killed Osama bin Laden. Other shadow or stealth wars are ongoing against Al Qaeda cells that operate training bases in Yemen, Somalia, Kenya, Pakistan, or former Soviet republics; these can feature private contractors and local operatives working with state intelligence agencies to locate key enemy commanders (Shane, Mazzetti and Worth, 2010).

Stealth and other wars can beget conflicts elsewhere in which ordinary people feature. An errant US Predator strikes a party of suspected Al Qaeda agents in the remote desert of Yemen. It kills five people, including a friendly provincial governor who was mediating between militants and the Yemeni government. Some months later, average Yemenis take to the streets, some with arms, to rid themselves of a loathed dictator. They become one national group among many in what will be the lingering Arab Spring of 2011, a time of mass uprisings against determined autocrats. Most such groups of citizens will fight their own states largely on their own. In the case of Libya, however, British and French planes will attack targets in Tripoli and elsewhere in support of anti-government militants fighting a sophisticated Libyan war machine.

Meanwhile, collective violence flares regularly between Israel and ordinary Palestinians. In Sierra Leone, Congo, and Liberia, packs of child soldiers ravaged rural villages for causes that are unclear; meanwhile, their soldier overseers engaged in mass rape as a war-fighting strategy or opportunistic crime; the Democratic Republic of Congo (DRC), where war has been ongoing for as long as many people can remember, is considered the rape capital of the world. A violent Maoist movement of mostly ordinary people conducts violence in India, and there is a recent history of anti-government guerilla movements in Mexico and Peru, Chechnya, Kashmir, and Thailand. Criminal gangs overrun northern Mexico, turning the border city of Juarez into a war zone. Two Korean states face each other across a cruelly misnamed demilitarized zone, as they have since the 1950s, despite a succession of peace talks over the years. Thai soldiers battle Cambodian forces around a string of ancient temples at their borderlands, which both countries seem willing to damage in a standoff over ownership.

Queries and concepts

Given a crowded picture of contemporary post-World War II wars, with their many centers of authority and agency – and I have only touched its edges – the main war questions to pose and address here are these: What is war and what is war about today? How does IR generally analyze war; that is to say, what does IR usually include and exclude in its war studies? How do related fields, as well as its own subfield of feminist IR, analyze war? What do they include and exclude? Can and should traditions of analysis be bridged in order to provide a fuller picture of war than any tradition alone has achieved? How? And how far can IR stray from the social sciences in order to study war as though people counted?

My approach to laying out the issues is to start simply and contingently, so that a variety of frameworks that are seemingly at odds can be read with, rather than against, each other. I look for nuggets of insight on war as bodily and emotional experience in IR and feminist texts that exemplify important traditions of thought, adding on related academic fields and literary sources. As various traditions are considered – from realist IR to feminist studies of war mourning, to novels featuring bodies, emotions, and institutions – the meanings of war and the experiences of war broaden, becoming more complicated rather than more parsimonious as the book proceeds. I use an exemplary text methodology, which enables me to juxtapose competing frames of analysis in pursuit of new ways of thinking about an old and very disciplining social institution. What results is an example of the art-making logic of collage, about which I have written (Sylvester, 2009, 2005). The approaches touch at points, leaving it to me as one analyst, and to readers with many perspectives, the task of pondering the ways, means, and subjects of war as so many pieces tacked to a board, whose interconnections must be theorized rather than assumed.

As a starting definition of war, *think of collective violence used to achieve a political agenda, the usual IR understanding of war; but think about the nature of that collective violence instead of its existence as a "mere" defining fact.* I will argue in the chapters ahead that war is a politics of injury: everything about war aims to injure people and/or their social surroundings as a way of resolving disagreement or, in some cases, encouraging disagreement if it is profitable to do so. As part of that mission, many will endeavor to protect themselves from injuries by fleeing the war zone, donning protective clothing, hiding, or looking away from war scenes on the television news; some people will be in a position both to injure and to evade injury at different moments in a war. The point is that injury is the content of war not the consequence of it, as Elaine Scarry (1985: 67), an analyst in the humanities not IR, was able to see a number of years ago. By treating war at a higher level of analysis, focusing very often on causes and correlates of war, on war strategies, weapons systems, national security interests and the like, IR repeatedly makes injury into a lamentable and regrettable consequence of the "normal" violence of war (e.g., Levy and Thompson, 2011; Vasquez, 2009).

Add two provisos. One is that all wars fought today have international components. Those components might be weapons recycled from earlier international wars (as in Afghanistan) or combatants who receive military training in one country to carry out acts of war in another (as Al Qaeda bases in Yemen and Pakistan). The provision of funds or military equipment to start or sustain a war counts as an international component (Libya), as do activities that breach international laws on war, genocide or human rights (Rwanda); outright attack of one state by other states, as in Iraq today, is also clearly an international intervention. The second proviso is that war should be studied as a social institution, which means that a range of people and their experiences and relations constitute and change war as much, perhaps, as developments in law, weaponry and strategy. War is an institution in the sense that heterosexuality or marriage is a social institution. Using Foucauldian terminology, such institutions have no fixed locations or administrators. Rather they are regimes of truth that emerge over time and dominate alternative ways of living to such a degree that they seem normal and natural, or at least unavoidable. In the case of war, the institutional components include: heroic myths and stories about battles for freedom and tragic losses; memories of war passed from generation to generation; the workings of defense departments and militaries; the production of war-accepting or -glorifying masculinities; the steady production and development of weapon systems; religions that continue to weigh issues of just and unjust wars instead of advocating no wars; and aspects of global popular culture – films, video games, TV shows, advertisements, pop songs, and fashion design – that tacitly support activities of violent politics by mimicking or modeling their elements in everyday circumstances.

When we are studying war as a social institution, people count as participants who have experiences and agencies related to war. It is not just the "important" people who affect and are affected by war, the ones whose photos appear in the media. *Everyday people are involved in the social institution of war in straightforward as well as complicated and often unnoted ways* – as combatants, yes, but also as mourners, protesters, enthusiasts, computer specialists, medical personnel, weapons designers, artists, novelists, journalists, refugees, parents, clergy, child soldiers, and school children. Globalization has redoubled the ways that everyday people can access war as members of private security companies, mercenaries, relief workers, photographers, and students. One can read, watch, or do war by logging onto YouTube and Facebook; or people can listen to the village radio for reports of local and distant wars. Trite though it is to say, few people are genuinely isolated and unaffected by specific wars or the constancy of war-supporting economies, war language, and war images. We are all part of what Vivienne Jabri (2006) refers to as a system matrix of war, which I call the transhistorical and transcultural social institution of war in its various particularities. In this book, the ordinary person alone or in a group is the essential unit, subject, and level of war analysis, and such persons can figure as sufferers of war, enthusiasts, and spectators of close and distant violent politics. In this global time, everyone is in that war matrix, which means that everyone has war experiences.

People can be inside the matrix in a variety of experiential ways. But how to study war as experience? The very word "experience" is commonplace – we all have experiences – and yet it is simultaneously abstract and difficult to characterize. In this study, it is taken as axiomatic that *war is experienced through the body, a unit that has agency to target and injure others in war and is also a target of war's capabilities.* The body is a biopolitical fact of war that we must not train ourselves any longer to avoid studying, or get so confused about that we cannot study "it." The body is also a contested and diverse entity that comes with gender, race, class, generational, cultural, and locational markings that affect and are affected by social experiences. Analysts from a wide array of disciplines raise questions concerning the body as a physical and material entity, a performative entity or a mostly symbolic or externally manipulated actor. Bodies can be thought of as biological, as cyborgs, as rampantly undead monsters or even as spaces filled with nonalive actants operating in assemblages (Bennett, 2010). Are bodies surfaces only or depths? Is the mind part of the body or something separate from it? In this book, the human body as a material and discrete entity emerges from the great stable of intriguing possibilities of its nature and existence. That the body might be a vampire, cyborg or robot does not matter at a general level; or rather, it is better to say that the many types and aspects of bodies that fascinate so many body theorists and cultural observers are socially mediated. Zombie culture would mediate zombie experience differently than the myriad specificities of culture that surrounding bodies that do not lead undead existences for up to 400 years.[2] The point is that bodies experience war in differentiated and mediated ways, but the body is central. *Experience is therefore the physical and emotional connections with war that people live – with their bodies and their minds and as social creatures in specific circumstances.* Experience as a salient concept and set of practices has been mostly neglected across IR and is only now being explored in the field's writings on war (Brighton, 2011; Sylvester, 2011a).[3]

The body can experience war physically – through wounds and attending to wounds, through running, firing, falling, having buildings fall on it, writing about war, filming moments of war, photographing war, feeling hungry or sick during war and so on. It can also experience war through emotions, a category that when pulled out and underscored can imply an unfortunate mind–body dualism. In 1997, feminist Elizabeth Spelman brusquely noted that "those of us drawn to Cartesian tidiness may be tempted to tote up our sufferings under two columns, one labeled 'mental' and one labeled 'physical' – though Descartes himself seemed to find this distinction confounded in the case of everyday pain" (Spelman, 1997: 4). Is the mind part of the body or something separate from it? It is *au courant* in much cultural studies literature, including feminist cultural studies, to deny – not dualism at all, but the role of the body as a source or even realistic location of emotions. The mind of the body, rather, is seen as the interpreter of feelings or the sensing center of affect, those psychological and physiological intensities that become emotions when they are given socially conditioned meaning. Accordingly, it could be said that many societies accept war or become inured to its violence and injurious nature, or buy into the threats and fears that governments project as justification for large militaries and their use, through emotions that line up with the prescribed war scripts.

Along with positing the importance of the body as the entity that experiences war, this study also takes a position in the discussions on affect and emotions and where these are lodged, arguing that the social determinist tendency in the study of emotions is undoubtedly not wrong; but it can be reductionist. Writing instead of reciprocities of body and mind, relays, and comminglings – giving the body some credit as the unit that senses and feels and thinks about its surroundings – is more sensible than saying that the body is out of the picture, to all intents and purposes, when it comes to emotional activities of all kinds, including war activities (Connolly, 2002). Despite writing about the body and war in "physical" and then "emotional" terms, there is an important sense in which the two actually interlock and mutually create experiences.[4]

Foregrounding IR and feminism

In a variety of contexts and ways over the past couple of years I have been asking for a war question (actually, a set of questions) to emerge in feminist International Relations thinking. By war question, I mean the development of an area of study – one could call it war studies – within academic feminism, as it is located in the wider field of IR. That call is decidedly not based on any personal belief in war as a good or usually even useful political strategy. The intent is not to glorify war. The intent is for feminist IR to study it with the same intensity that it has tackled other large social institutions like gender, motherhood, religion, marriage, and heterosexuality. War has not been studied with any comparable degree of systematic attention within the ranks of feminism, which is not to say that there is no work on war. In the following chapters I refer to numerous feminist tracts that explore aspects of war, including militarization, militarized masculinity, women in Western militaries, women suicide bombers or militants, and women and political violence. Still, because feminist thinking has historically pitched against war and other forms of violence, feminism within IR has also been comfortable studying peace rather than war – even though war is a core subject of IR. Put differently, war is not a social institution that feminism "owns;" rather, feminist analysis usually depicts war as someone else's misguided institution, someone else's area of study. The feminist mission is to end war as an institution and set of practices.

And that position makes sense. It makes sense to be against war as a politics, economy, sociology, and performance of power. It is rational to worry about the militarization of the world today and to fret about young American men and women sent each generation to foment or fight wars far from US shores, such as in Afghanistan, Iraq, Bosnia, and Vietnam. It makes sense, too, to worry about people who are killed in distant wars by Western troops and by local marauders. It makes sense to hate war's shameless destructiveness, repeated and repeated through the ages as though war were somehow utterly indispensable to political life. Indeed, if there is one social institution that is both historical and transcultural it is war. Intriguingly, however, feminism was also opposed for many years to organized religion, to marriage, to heteronormativity, yet studied those institutions deeply and systematically. That war is a relatively neglected topic of feminist analysis is odd, especially since war is gender-inclusive in its components and impacts. No one can claim to be outside the institution of war, although anyone can hold normative positions against it. Myself, I have no stomach for war, but war is just as much ours to investigate as anything that affects embodied women and "women" and "men" as subject statuses intersecting with other subject statuses. Having implicitly assigned war, however, to originary spaces tangential to our remit, feminist literatures pick and choose what to study on war. War philosophy, military history, war strategy, battlefield tactics, weapons and war industry analysis are generally not part of the narrow and spotty feminist focus on war.

Women and men who join the institution of war or work with it can also put feminists in a conundrum: to support them, not support them, take no position, look the other way? There's unease about khaki and how it can become some women (Enloe, 1983), just as there is unease about other women who say they are feminists but advocate positions that tend to fall outside the usual feminist narrative. Hirsi Ali is one such woman – a difficult feminist whose reception illustrates the way lines can be drawn and people thrust out and away from feminist analysis. Ali is the Somalian-Dutch feminist and former MP, who writes antagonistically about the effects of Islam on women adherents. Feminist thinking was once secularist but is now cautious about giving offense by supporting people and debate positions that could be interpreted as orientalist. Its positions on certain topics can accord with Michael Sandel's (1998) description of liberalism in Western countries today as empty at the center, having relinquished all positions in favor of an overarching tolerance. So Hirsi Ali, whose critique of a religion on feminist grounds would once have been embraced by feminism, ends up dangerously isolated with a security problem that requires bodyguards to keep at bay.

And here is another contradiction within that contradiction: feminism today honors difference. It oftentimes does so, however, by assessing how local communities of feminists and nonfeminists are responding to controversial issues about women and rights. The difference approach corrects the tendency in the

1960s and 1970s generation of feminism to use the term "women" in well-intentioned but nonetheless universalized ways, as in (all) women are oppressed by patriarchy, capitalism, biased public policies and traditions, bad research methods, and so on. All women were, in fact, often Western women speaking of ourselves and projecting Western feminist thinking outwards on cultural others. The necessary corrective was that feminists started heeding the experiences of others by supporting women whose lived experiences entailed making choices that we might not think are choices at all. We became quieter about our own experiences and less sure that we could know what is right for women whose daily lives are different than our own. Difference thinking is reasonable and practical, a strong approach for a multicultural, multiracial, and multiclass world.

It can seem, however, that upholding difference in order to give others voice can become an end in itself rather than a means of determining feminist positions. In some cases, feminist positions can be proclaimed if they accord with local views imprimatured by groups living in the situations under discussion – living as Muslims, living in societies that practice female circumcision or that tolerate sex work. When a self-proclaimed feminist takes up a difficult or unpopular political position today in the name of feminism, problems arise. Feminist Hirsi Ali (2006) criticizes Islam. The Iranian feminist Azar Nafisi (2004) secretly reads and analyzes Nabokov's novel, *Lolita*, which is favorable toward sex with underage girls, with women university students during the Iranian revolution. Both women have defied local cultural authority on the grounds that it cripples women's agency. Yet such women are differences too – within feminism and with respect to culturally approved modes of behavior. Have they crossed an invisible line to become a difference that is so far beyond difference that it cannot be handled within the framework of difference feminism? A similar problem can result when some feminist analysts choose to investigate aspects of war rather than focus on peace, when they treat warring women's actions and politics as instances of agency instead of unacceptable exceptions to the peace position. At such times, feminist academics can become knowledge gatekeepers, excluding rather than including controversial positions. This was something we decried when feminist positions were vetted and found wanting by traditional academic fields. Is the correct position on difficult issues and people the one that a majority of feminists or publics holds, even if those views are unmoored from the concerns with tolerance and cultural specificity that gave rise to difference feminist thinking? I think not. In a complicated world, it is important to include "difficult" people and topics and differences, rather than decide a priori who/what is in and who/what is out.

IR has its internal progression and legendary discipline-defining debates, too. As a central tendency, however, it does not worry too much about people's experiences of international relations. War was one of the instigating politics behind the emergence of IR in the UK on the heels of World War I. IR is not necessarily pro-war, although some have argued that is complicit in the many violences of our time (Smith, 2004). Rather, the field has historically viewed war as one of the political and material dynamics that emerge when nation-states and other actors operate beyond the realm of enforceable rules of law that characterize domestic politics in many (ideal) states. IR considers factors behind historic wars involving international actors, the frequency and types of wars, the weapons and strategies of war, public responses to wars, and actions that can work to quell war: all of this has a place in IR's stable of knowledge. As the international system and the world change, however, many of IR's traditional ways of understanding the sources of war, combatants, and the way war is waged and resolved, must also change. I argue in this book that it is time to add people to IR's stable of war actors and not exclude or marginalize them any more. Just as some feminists find that what they once advocated is now *passé*, so too must IR face up to their bits of obsolescent thinking. The state system theorized by the field is no longer the sole keeper of international war; nor is war what it was at the time of World War II or during the Cold War period.

The social institution of war and the range of people's involvement with it is what many in feminist IR believe that IR should study *alongside* its other categories and levels of analysis. Indeed, the goal should be to link levels in ways that provide a bigger, more complete picture of a phenomenon, a picture that cannot as easily be reduced to acronyms and to impersonal language overall. IR has not gotten quite that far yet, even though the field has evolved a capacious camp structure in recent years that accommodates every topic, methodological approach, and constituency that clamors to be part of IR (Sylvester, 2007). Each of about 30 camps has its favorite personages, texts, journals and identity codas, for example, there are a lot of women in the feminist IR camp and a lot of Europeans in the international political sociology camp. The camp structure of IR emerged rather quickly when the Cold War ended and parts of the world once pronounced on from a distance turned up in IR and began influencing its agendas. Simultaneously, a "third debate" on epistemology worked its way through IR and raised probing critiques that fueled disciplinary chagrin when IR failed to anticipate the fall of a Cold War wall. People – everyday people, masses of defiant people – took to the streets in East Berlin and simply, eloquently broke boundary rules and lived through the experience. Such people were not even a glimmer in the eye of reigning IR theories of the time.

They still are not: IR was taken by surprise again when the same kinds of people rose against the autocrats of one after another Middle Eastern state in 2011.[5] The field seems hogtied to the notion that ordinary people cannot have the type of agency in international relations that can shift polarity or overpower embedded authoritarian regimes. But at least today's IR shelters more people and places of the international than it did 30 years ago, and investigates a wider set of relations than those associated with the (always big G and big P) Great Powers. It also disperses disciplinary power to the point that no one camp can define the field or rid IR of any "undesirable" analytic tendency. There is no realm of heresy in democratized IR. Yet there are still many blind spots. For one, camp dispersals can make intersectional analysis difficult. Billowing smoke from 30 or so campfires obscures commonalities and promotes intellectual identity politics instead of mutual learning. If one is in the realist camp, the emphasis on states as actors and national security logics of military action abroad remains. If, instead, one camps with constructivists, then norms around collective violence might be the way to think about war. Camps associated with critical approaches to IR – poststructuralism, postcolonial analysis, political sociology, critical security studies, critical studies of terrorism, or feminist IR – share skepticism of inherited IR knowledge as limited or wrong-minded, owing to the near-hypnotic hold that positivist parsimonies and either-or categories have on the IR imagination; tellingly, few of the new camps are in tow to positivism. Within the large critical arenas of IR today, though, camp boundaries are protected and new participants can be informally judged by the group as to their suitability: do they use the right sources and agree with the main tenets of the camp?

The one area of IR where intersectionality reigns is security studies (Buzan and Hansen, 2009). Security studies assumes that war is possible but casts its many, bee-like eyes on conditions that threaten the well-being of states and people. It is also said that security research keeps one eye on policy, which means it often presents an instrumentalized account of war as something to prevail within or gain security from (Brighton, 2011: 102). At the same time, IR's many camps are places of academic debate, and "What does 'security' versus 'war' mean?" can elicit multiple answers between and within them. Lene Hansen, a leading figure in contemporary security studies, suggests that there is no hard and fast division between war studies and security studies, since war remains the ontological foundation of security studies. The danger or possibility that force will be used in a situation is what makes "security" a particular form of politics.[6] Still, there does seem to be some disinclination within contemporary IR to study war with the same energy that is now devoted to studying security. It is a problem that parallels feminist IR's disinclination until recently to break out of the peace niche of feminism to study war as a social institution.

The feminist IR camp does have a vibrant security wing; and it has developed a war studies wing, too. In fact, the two interests and groups of analysts crisscross each other. Peace-studies and war-studies feminists share concerns about war, whereas analysts who focus more on security links with environmental, health, or migration issues usually have a different starting point.[7] In any event, the new generation of thinking in feminist IR angles into war studies without apology and without forsaking feminism (e.g., Sjoberg and Gentry, 2007; Wibben, 2011; MacKenzie, 2009; Parashar, 2009; Sylvester, 2005; Sylvester and Parashar, 2009). It studies women who engage in war violence or who encourage others to do so, as well as those who are victims of gender-based violence in war. The focus is directly on people and their experiences with war, and those people are often researched in situ using interviews and other ethnographic methodologies usually associated with anthropology. Several of these works and methodological approaches will appear in the following chapters. Suffice it to say at this moment that whereas experience often surfaces in feminist analysis as positionality or standpoint, it is conceptualized more broadly in this book as a reflection of new feminist IR writings that take up physical and emotional aspects of war.

The layout

War as Experience unfolds in the following way. Chapters 1 and 2 continue to introduce and discuss the components of IR's war interest by setting up a call-and-response dynamic across four IR camps and the new war studies wing of feminist IR. Chapter 1 overviews war thinking in realism, new war studies, constructivism, and critical IR. These are obviously not the only locations of war analysis, but they are instructive for the purposes of revealing the kinds of queries about war that emerge in IR. Rather than try to summarize all IR literatures on war, a task that others have undertaken (e.g. see Lebow, 2010), the emphasis is on individual exemplars of each tradition. The realist camp is represented by John Mearsheimer and Stephen Walt; the new wars camp by Mary Kaldor; and critical IR by Tarak Barkawi, Shane Brighton, and James Der Derian. There are also two exemplary crossover texts discussed: one by Michael Shapiro combines post-structuralism with concerns that are usually associated with feminist IR; the other is by Karen Fierke who works within the constructivist tradition to address links between individual, group, and national war traumas. Each exemplary text addresses some aspect of contemporary wars and warring in camp-characteristic ways. At the same time each also, sometimes inadvertently, reveals more than abstractions of war that might preoccupy their colleagues. People lurk as shadows and ghosts in all the selected texts. It is important to give those people greater visibility and yet also credit IR with harboring them, as one step to communicating across camp traditions.

Chapter 2 presents the new focus in (some) feminist IR on studying war, and studying it in ways that depart from usual IR approaches. It considers original feminist presentations of war and bodies by Jean Bethke Elshtain and Cynthia Enloe and exemplary contemporary texts by Enloe, Miranda Alison, Megan MacKenzie, and Annick Wibben. Two of the latter use ethnographic methodologies in their research (Alison and MacKenzie), and the third text by Wibben presents narrative and discourse methodologies for pondering issues of war and security. Chapter 2 also highlights the long feminist tradition of basing knowledge on concrete bodily experiences of women, an approach that both marks the existence of feminist thought and unleashes internal debates and rejoinders. The chapter closes with a discussion of moments of unexpected overlaps of feminist IR and IR scholarship of other types around issues of people in war.

Chapter 3 marks a shift in focus from literatures of war to reconsidered concepts, starting with the body. As noted above, there is considerable debate within and outside of feminist IR on what the body is and whether there is a split between body as flesh and the mind of emotions and thought. This chapter highlights the insightful contribution Elaine Scarry (1985) has made to the study of war by presenting bodily injuring as the content of war and not just its unfortunate consequence. It then asks what is injured in which ways in war, concentrating on the physical body as one cornerstone element of feminist theorizing on gender, society, and women. Other than Scarry's exemplary text, the chapter foregrounds a range of works that reveal the complexities of thinking, feminist and not, about the body. Included are discussions of works by Veena Das, Judith Butler, Joshua Goldstein, Donna Haraway, Erin Manning, Daniel Drezner, Giorgio Agamben, and Jewell Gomez. Two exemplary texts, by Dubravka Zarkov and by co-investigators Maria Eriksson Baaz and Maria Stern, illustrate feminist-inspired research on an increasingly important topic in war studies overall: rape.

In addition, this chapter shifts ground by introducing fictional works and testimonial accounts of war experiences as exemplary war sources. The two initial pieces discussed are by Tracy Kidder (2010) and Gil Courtemanche (2003). The first presents one man's harrowing "true" physical experiences entailed in escaping the Burundi-Rwanda genocides, and the second offers a "fictional" character who makes the choice to die from AIDs rather than suffer the chop of a neighbor's machete in Rwanda in 1994. Personal narratives of war are central to the feminist IR analyses that are discussed in Chapter 2. Kidder's and Courtemanche's extended portrayals of victims and survivors of "new wars" set a context, put the reader into it, and spark interest in learning more by raising large war issues as matters of personal experience. As for using fiction in IR more generally, I have argued (Sylvester, 2000) that the line between fact and fiction can be very thin, so thin that one cannot see it clearly in certain circumstances. Novels set in postcolonial locations of actual war, for instance, reveal aspects of reality that can be inaccessible to the Western social science researcher, such as the possible thinking of killers, victims and spectators; the actions each takes to survive; and the sense of urgency and fear as well as the joys and calculated behaviors that surround war zones. A given story might be hypothetical in certain elements, but it cannot be too much so and also satisfy informed readers that the events described are entirely reasonable for the context.

The acclaimed Zimbabwean writer, Chenjerai Hove (1994: 15) goes further. He does not even think that one should worry about a close fit between reality and fiction, in any event: "people themselves are bits of imagination. We are invented. We are invented by other people." In a social networking era, the "truth" of who someone is and the "fictions" that a person might send out over cyberspace can be difficult to differentiate; and yet these can figure prominently in that person's job and other social prospects. Today's treasured facts can also become tomorrow's fictions, as knowledge in all areas is updated and revised. Perhaps the most compelling justification for including fiction in this chapter and those that follow, however, is the one Hove, writing with Okey Ndibe (Ndibe and Hove 2009: 11), has given more recently: "The idea of pure art, an art uninfected or uninflected by the surrounds of political upheavals and pervasive social misery, is a myth. Art is shaped by, and shapes, all facets of experience."

Chapter 4 focuses on war as experience of an emotional nature. It is a discussion that considers IR's peculiar fascination with rationality as the *sine qua non* of human capacity and central element of state and statesmen behaviors. Against that strong view, which is held by many US-based scholars of IR, research by neuroscientists attests that rationality does not exist in the absence of emotion:

in cases where pure rationality is observed, the subject is likely to be suffering mental health issues. Several students of IR make the point that emotions must have a more prominent place in that field, but two pieces are the focus here: one by Neta Crawford (2000) and the more recent Forum on Emotion and the Feminist IR Researcher in *International Studies Review* in 2011 (Sylvester, 2011b). Crawford's (2000:125) *definition of emotions takes the lead in this study as "the inner states that individuals describe to others as feelings, and those feelings may be associated with biological, cognitive, and behavioral states and changes. Feelings are internally experienced, but the meaning attached to those feelings, the behaviors associated with them, and the recognition of emotions in others are cognitively and culturally construed and constructed."* This definition is in line with much research in neuroscience as well as additional thinking by a range of researchers mentioned in the chapter, including the feminist theorist Lauren Berlant and political theorists William Connolly and Brian Massumi. The chapter concludes with a consideration of Judith Butler's sophisticated feminist ideas on the range of emotions that can connect people internationally and politically around grief over war's losses, and places her work in a crafted dialogue with Erin Manning and Berlant that suggests ways to deepen theorizing in this area. Chapter 4 also brings in the emotional contradictions of being both sufferer and spectator of war as depicted in Chimamanda Ngozi Adichie's novel, *Half a Yellow Sun* (2006).

The final chapter reminds us where we have been in the book and why it has been important to take a journey to war that highlights the centrality of people to a longstanding social institution. Indeed, people and war become so intertwined here, in ways often anticipated by feminist thinking but not by IR, that war becomes, in effect, a realm mostly of experience rather than a set of causes and correlates and abstract actors. A point worth emphasizing is that while much research relevant to issues of war and people is associated with feminist IR and broader feminist traditions, all traditions of analysis have knowledges to share and to reconfigure. The idea is to move toward analyses of war that can be like moving collages of juxtaposed parts rather than the law-like theories that were favored by IR in the past. Experiences of war exist in horrid gore and in grey zones of war, peace, and spaces of the personal, the local and the international that can deny contemporary war studies any ease in locating and characterizing war's "true" sources and power. It is fertile ground for a number of IR traditions to collaborate around, such as constructivism and political sociology on war as a social institution, feminist and critical IR on experience, and critical and neuropolitical IR on the body and emotions. War is also a topic prime for the inclusion of insights from literature, history, anthropology, sociology, the visual arts, postcolonial studies, philosophy, and a raft of other fields – precisely because no group is outside war in our global time.

That is a key point: we are all touched very directly by war through spectatorship and/or by war's chilling violence. Since war is an institution that is continuously shape-shifting, it is time that IR worked across camp and field barricades to study war as experience. To repeat my challenge to IR war studies: study up and study down as a linked project, rather than favoring elites and presumed power centers of war over "collaterals" who also experience war but are kept out of view, assigned to some other field to study. Those of us who are associated with IR in some manner must stop averting our eyes and decide to descend into the ordinary of violence, as the feminist anthropologist Veena Das (2007) puts it, where people of international relations dwell, experience and can and do shape outcomes of considerable importance.

PART I

International Relations and feminists consider war

1

IR TAKES ON WAR

How should and how does IR study war? Notwithstanding the recent security studies trend in IR, much of the field, particularly North American IR, has been concerned historically with the causes of war and its general manifestations. It has not been especially interested in war as a social institution or war as involving physical and emotional experiences of collective violence. IR is at ease with highly abstracted realities, such as the state, militaries, norms, ideologies, international organizations, resources, rationality, decision-makers, and so on. When dealing with war, IR also has a penchant for matching conceptual abstractions with system-level dynamics that embroil states but not humans in wars – anarchy, polarity, power transitions, security dilemmas, offensive and defensive strategic considerations and the like. To the degree that they are brought into IR at all, people's experiences are coded as casualties or collateral damage and are seen as the consequence of war but not the main point – unless your name was Osama bin Laden, whose death in 2011 was very much one of the points of the war on terror. Similarly, IR is not inclined to see war as a social institution so much as a strategy of political violence available to states and other actors to resolve intransigent political problems or to augment state power and prestige. The closest it gets to a notion of an institution that is transhistorical and transcultural is the realist sense that international political systems lacking central governance, like the Westphalian states system dating to the middle of the seventeenth century, will prepare for war and use war to attain certain of their goals.

For most of IR's history, there has been no everydayness to the international, no ordinary people, no mundane social support structure for the big actions of states other than public opinion. War is heroics and tragedies, but few people have been studied by IR doing heroic or tragic things except statesmen and their nemeses. In the following sections, we consider a few exemplars of IR war thinking among leading scholars of today's wars: realism and constructivism in the United States, new war thinking in Europe, and critical IR analysis, which spans the Atlantic and Pacific but is mostly associated with European approaches to IR. Feminist IR fits into the third category – critical analysis – and is treated in detail in the next chapter.

Empty canvases of Western warring? John Mearsheimer and Stephen Walt

The most longstanding theoretical approach to war in IR is associated with various realist traditions of scholarship, of which John Mearsheimer and Stephen Walt are two of its leading proponents. Realism, as Mearsheimer (1994/95: 9) himself puts it, "paints a rather grim picture of world politics." It emphasizes the security interests of states operating as rational actors in a system of formal political anarchy, which is to say in the absence of an overarching government. State survival is all, which means "daily life is a struggle for power, where each state strives not only to be the most powerful actor in the system, but also to ensure that no other state achieves that lofty position" (p. 9). Left largely to their own devices, states use their offensive military capabilities to enact the tragedy of war in order to attain their security goals; along the way they might succumb to what earlier classical realism posited as the evils inherent in human nature to threaten others, or stumble because they have not acted in line with rational, instrumental decision-making required to manage threats (Wolfers, 1962; Lebow, 2010). In each version of realism, constraints on unlimited war exist at the system level, the main one being the balance of state power that forms and re-forms in an anarchic system. Relations across the system can be tense, however, owing to the perceived need to be vigilant in maintaining power through alliances and military preparedness. Arms races and security dilemmas ensue (Herz, 2003), but balances often prevent all-out interstate war. Still, sustained peace is largely a chimera in such a system.

The abrupt end of the Cold War bipolar balance of power – without a war and with a superpower voluntarily relinquishing its authority – put many realist assumptions under intense and skeptical scrutiny. So many state-challenging acts were carried out by the wrong actors according to the theory – people driving Trabant cars through the Berlin Wall, mothers pushing baby carriages through Checkpoint Charlie. The subsequent crumble of the Soviet Union and its power bloc, which realist theory would not anticipate, owing to its certainty about states striving for power or survival, could have ruined realist theory. The real events just did not conform to realism's basic tenets. The tradition has endured, however, by maneuvering notably around its exposed weaknesses. Some theorists have acknowledged that internal factors were important in bloc disarticulation; yet argue overall that external pressure over the years had softened up the system. Others claimed that the Soviet effort to match and exceed Western military technologies precipitated a collapse that was realized through popular resistance (Deudney and Ikenberry, 1991/92; Lebow and Risse-Kappan, 1997). Briefly and powerfully on the international scene as important actors, "people" are rapidly pushed to the margins of the realist analytic picture of the times, and then out of that picture entirely. When the conflicts and wars of the post-Cold War period started almost immediately, with no time for any peace dividend to materialize, the basic realist framework was applied to regional and sub-state flare-ups of collective violence (Posen, 1993; Van Evera, 1994).

An ensemble of pushes and pulls on states today can give rise to foreign policies on conflicts in the world that alarm Mearsheimer and Walt, the two most influential contemporary realists of IR. Each addresses such concerns in his own work and together as co-authors (Mearsheimer, 2001, 1990; Mearsheimer and Walt, 2007). Mearsheimer argues that great powers still matter, still strive to accrue power relative to other states, and will still go to war when necessary to prevent another great power from achieving hegemony. Despite the collapse of the Soviet Union, the structure of the system remains a formal anarchy, which means that the older saga of relentless security competition continues, albeit in new forms, such as the wars that emerged in former Soviet imperial areas and in the erstwhile Yugoslavia. The change to unipolarity or loose multipolarity under US leadership, and to a more thoroughgoing system of globalization, was destabilizing as well as liberating. But the alliance structures of the West creakily persevere and states continue to prize security and to calculate the consequences of going to war relative to gaining desirable outcomes through other means. The involvement of the US in wars in Iraq and Afghanistan, however, as in Vietnam earlier, has given these realists pause: they seem to offer relatively less in gains than in costs, especially in the case of Iraq. Mearsheimer and Walt explain American involvement in the second Iraq War starting in 2003 as a breakdown in foreign policy decision-making rationality, with the clearest general statement articulated by Walt in a brief *Foreign Policy* (2011) article. Charging that the US tends to war first and then think about it, he points to persistent irrationalities informing American war decisions: its belief that the US is an exceptional nation-state without peer competitors, that it must justify its all-volunteer military with periodic action, the lingering neoconservative and liberal interventionist influence in Washington, and a steady accrual of war powers in the executive branch. Notwithstanding the many types of rationality that have been discussed in other IR works (e.g., Tickner, 1988), what any rational interstate war is like from experiential points of view, relative to irrational wars, is flatly absent from the realist concerns with dangerous, irrational outcomes. Such considerations lie outside realist theory.

There are other tensions in the determination to show that failures of rationality can explain mistaken war ventures in our time. Realism's law-like propositions about international relations study down from a proverbial God's-eye view. The problem is that the small fry of the world often pay little attention to realist theories and can entangle others in locally-based logics of violence and war. In their jointly authored and highly controversial article, "The Israel Lobby and U.S. Foreign Policy" (2006), and subsequent book of the same name, Mearsheimer and Walt argue that US foreign policy toward Israel is irrationally favorable, to the point of giving Israel nearly unconditional support. The US consistently accords Israel "substantial diplomatic, economic, and military support even when Israel takes actions the United States opposes," such as building Israeli settlements in the West Bank and Gaza, or using cluster bombs in the 2006 Lebanon war (Mearsheimer and Walt, 2009: 64). Blind support (although there are always disagreements between the two states) is not rational in the sense that rationality is used in realist theory, roughly in cost–benefit terms. The problem, as Mearsheimer and Walt see it, is that a powerful Israel lobby within the US exerts pressure on national decision-makers to support Israel at all times, discouraging any type of support that is contingent on calculations of probable threats to Israel's or US security. Support, come what may, is harmful in their view to both countries in the post-9/11 period. In keeping with realist thinking, the co-authors urge the US to apply more strategic or even moral logic to their decision-making around Israel.

From a perspective on war as experience, it is intriguing to find that when presenting their main arguments about US–Israel relations, nary an ordinary American, Israeli, or Palestinian comes into clear sight; meanwhile, prominent individuals associated with the Israel lobby are mentioned by name. Very elliptically, however, Mearsheimer and Walt suggest that specific conflicts associated with tense Israeli–Palestinian relations, conflicts that include wars, do involve people and not just states. The experiences of ordinary people appear as ghosts fleetingly haunting the abstract discussion whenever the authors want to strengthen or illustrate their points in order to render them more persuasive and convincing. Across the text, they refer to Arabs, Jews, Jewish-Americans, the Palestinian people, Muslims, Palestinian protestors, neighbors, academics, professors, columnists, torture, massacres, rapes, executions, prisoners of war, civilians, beaten children, bystanders, students, peace activists, a 23-year-old American crushed by an Israeli bulldozer, and soldiers. Here is an example. Speaking of the first intifada of 1987–91, the authors note that

> the IDF distributed truncheons to its troops and encouraged them to break the bones of Palestinian protestors. The Swedish "Save the Children" organization estimated that "23,600 to 29,900 children required medical treatment for their beating injuries in the first two years of the intifada," with nearly one-third sustaining broken bones. It also estimated that "nearly one-third of the beaten children were aged ten and under." Israel's response to the second intifada (2000–05) has been even more violent … Israel has killed 3.4 Palestinians for every Israeli lost, the majority of whom have been innocent bystanders; the ratio of Palestinian to Israeli children killed is even higher (5.7 to one). (P. 38)

Agree with their overall argument or not[1]: the point for this discussion is that realists who would not blink when claiming that states war, refer at least obliquely to people's actual war experiences when they critique state policies that make no sense on usual realist grounds. They do so in order to justify their argument that Israel does not necessarily deserve the unconditional support that the US gives because it fails to act rationally or morally in its conflictual relations with Palestinians. In other words, the authors tacitly show that abstract arguments about war and foreign policy do not on their own blast out enough power to persuade decision-makers that they should alter their course. Although most people in the conflicts Mearsheimer and Walt describe are then subsumed by the argument that the US is repeatedly willing to sacrifice or suspend its own security interests for the "good" of Israel, their actual presence in the text, and the reasons for that presence, cannot be ignored. Even realism recognizes war as injurious experience, among the many other things it is. Walt (2009, unpaginated): "When you kill tens of thousands of people in other countries – and sometimes for no good reason – you shouldn't be surprised when people in those countries are enraged by this behavior and interested in revenge."

Both of these realists put little faith in any war-prevention capacity of international institutions. Mearsheimer argues that such IOs, including the main peace institution of our time, the UN, have been created by states and are beholden to them. While constructivism and neoliberal institutionalism talk about norm creation and habits of reciprocity that have to do with state interactions through institutions and international law, Mearsheimer (1994–5: 7) categorically asserts that these political forms "have no independent effect on state behavior." This is a difficult argument to sustain in light of the UN Security Council justification for supporting a military fly-over operation in Libya in 2011 on the grounds of the 2009 General Assembly Resolution on the Responsibility to Protect. One could argue that it has been in numerous states' hard-nosed interest to see Qaddafi exited from Libya. But Anne Orford, a leading feminist analyst of humanitarian law, calls the UN turn towards Responsibility to Protect revolutionary. She says it replaces the usual functionalist-technical sense that the most capable and willing *state* should intervene to protect populations, rendering authority to determine responsibility to protect "merely" a routine practice of international organization.[2] Beyond that particular concern, it is difficult to argue that human rights more generally is not a norm and set of practices installed in global organizations and discourse. Humanitarianism and older sovereignty-upholding norms and security concerns coexist, compete, or run along parallel tracks in today's international relations,[3] which suggests that people, rights, and organizations will appear in realist analysis here and there. They cannot be written out even if one tries.

The problem of acknowledging people in realist international relations is that realist IR is heavily invested in studying the causes of war and state behaviors as leading war forces. I am particularly fond of Mark Duffield's (2001: 13) poetic description of the obsessive search for wars' "causes and motives . . . [as] rather like Victorian butterfly collectors, [seeking] to construct lists and typologies of the different species identified." If Mearsheimer and Walt had elected, as I realize they would never do, to consider the problem of war as a clearly layered phenomenon that must extend to those people they mention only briefly, state rationalities and irrationalities would share space with rational and irrational experiences and aftermaths of all kinds – not just outcomes that threaten US and other state national interests. And those considerations might circle back to influence the humanitarian agenda of the times as well as state foreign policies that juggle security with responsibilities to protect.

Many new wars and experiences? Mary Kaldor

As the briefly feted concept of "peace dividend," and its more sustaining cousin, "democratic peace," suggested, the collapse of the German Democratic Republic and then the Soviet Union was interpreted by many in IR as the final end – of war if not exactly history. The major point Francis Fukuyama (1989) made in his stunningly simple piece on the end of history was that the major dialectical currents of international relations had run their course. The last major cleavage had resolved itself in favor of capitalist democracy and there were no comparable epochal battles left to fight. Yet war was not at any end, as we all know. Realists were forced by circumstances to apply their frameworks to post-Cold War conflicts orchestrated mostly within states rather than between them. So were others in IR who had been less bound into Cold War theorizing.

By the turn of the Millennium, the talk in many European IR circles was of new wars emerging within the opportunity structures created by the Soviet fade-out and the full-fledged globalization that followed. There were said to be several new elements of warring in the 1990s, such as the politicization of ethnicity and religion, strategies of outright genocide that included brutal civilian-on-civilian assaults or that targeted civilians, and networks of criminality that sustain wars by trading arms and drugs as well as looting museums and smuggling minerals. For many, the main characteristic of new wars that emerged in Bosnia, Kosovo, Iraq, Somalia, Rwanda, Liberia, Sierra Leone, Sri Lanka, and between Israel and the Palestinians in the 1990s was their seeming irrationality, in part or in toto. Not unlike the Mearsheimer/Walt stance on the centrality and importance of states acting as rational actors, it has seemed that these wars were carved out of spaces of emotion and manufactured animosity rather than through rational deliberation on war relative to other strategies for survival or power enhancement. "New wars" thinking continues, therefore, a tradition of thinking about rationality, but does so by foregrounding seemingly irrational elements in local dynamics of wars that might once have been called civil wars.

The "new wars" concept is closely associated with ideas Mary Kaldor has been discussing since 1999. Kaldor (2006) argues that wars might be declining in number in the post-Cold War period, but the scale of political violence has increased exponentially in the early twenty-first century. The change, she argues, has less to do with greatly enhanced military capacities in the information age – which are considerable – than with altered social relations of warfare. Globalization has brought about contradictory tendencies, such as the fragmentation of the Soviet empire alongside forces of global economic integration, the politicization of ethnic and religious identities earlier subordinated to a homogenizing national discourse or refused and punished under the old regimes, and the accompanying emphasis on local politics and local economies within the enlarged global political economy. But it is more than this, too. Globalization has made it easier than ever for people to know what is going on in international hot spots, leading, she would undoubtedly say, to the diffusion of ideas such as those the world witnessed in 2011 among reformist movements across the Middle East. It also makes it easier for arms, diamonds, antiquities, and mercenaries to be traded globally, both legally and illegally, and greatly enables "trade" across a development industry that crisscrosses poor countries. It is the kind of new order that features human rights and humanitarianism and also kill-to-be-kind logics behind wars of humanitarian intervention (Sylvester, 2005). It is a time period when many social relations can be monitored for security threats, including airplane travel, everyday life on the streets of London, one's computer files and overseas bank accounts.

Kaldor's interest lies mostly in understanding predatory wars of the globalized era. These occur in areas of weak state control, failed states, or disintegrating states, where one of the hallmarks of state authority, the legitimate control of armed violence, has been lost. In its place are outside militaries, as in Afghanistan from 2001 to the present, or private militaries of local and transnational composition that keep the peace and, as is more likely, ravage local areas. Indeed, Kaldor argues that the wars of our time might be reversing the processes associated with the rise of the nation-state. Instead of consolidating decentralized authorities and their militaries, and bringing greater peace and stability to a region, the tendency in areas of new wars is for centralized authority to break down and be replaced by criminal networks, multiple armed authorities, corrupt practices, mindless local barbarity, population expulsions, and the organized politics of fear instead of civil security. The armed groups can look like guerilla fighters but they act like gangs: their goal is not to win over the population to their vision of the future, but to destabilize local areas by terrorizing people through "the use of spectacular, often gruesome, violence to create fear and conflict (p. 9)." Considerable attention has been drawn to the use of rape as a weapon of new wars and the conscription of children to war fighting (e.g., Nordstrom, 1997; Eriksson Baaz and Stern, 2010). But many new wars also make use of advanced military technologies, such as lighter rifles and small but powerful landmines.

Missing from all this is what Kaldor (1999: 7) refers to as a "forward-looking project – ideas about how society should be organized." So, for example, the national liberation wars of Africa were waged for political independence under a particular ideological framework. Today, some of those guiding frameworks, particularly those associated with communism, have been discredited and/or subverted by those postcolonial leaders who have embraced de-development or continued underdevelopment as a governance strategy (e.g., in Zimbabwe, Congo, Uganda, Somalia, pre-2011 Egypt), while personally reaping the rewards of transnational criminal activities around minerals, looting, human trafficking and applying international humanitarian aid to the war effort. Kaldor refers to these activities as signs of "a war logic built into the functioning of the economy" (p. 10); there is little incentive to end the lucrative disorder to take up legitimate commerce. One thinks of the hash poppies of Afghanistan that keep the Taliban well financed. Blood diamonds kept Charles Taylor thriving in Liberia and they allegedly maintain Robert Mugabe's military and extended family in fine form as the rest of the population suffers unemployment, under-nutrition, diseases, and deteriorating infrastructure (albeit not outright war ... yet). And then there are the kidnappings for ransom by radical Islamist groups. To Kaldor, the nature of the violence is a key indicator of the new war: all sides participate in the suppression of normal civility and operate effectively on the undersides of morality and cosmopolitanism. That is to say, the new social relations of war are, in her terms, "retrograde" (p. 10). To counter them, Kaldor (p. 11) advocates the types of local, national, and international processes that advance a "forward-looking cosmopolitan political project which would cross the global/local divide and reconstruct legitimacy around an inclusive, democratic set of values ... to be counterposed against the politics of [identity-based] exclusivism." Such an approach, she argues, would leave behind the draconian methods of economic structural adjustment on the one hand and liberal humanitarianism on the other. For her, international law on human rights and war conduct needs to be reflected in institutions that seek to repair the social relations of war and incivility.

New war thinking is part and parcel of European IR, where institutionalist and sociological literatures are deemed more appropriate to bring to bear on international relations than is common in the USA. Europeans are also more physically proximate to many new war zones and have colonial experiences that can illuminate social relations and histories of war. Yet new war thinking is also repeatedly critiqued by European IR scholars as old warring in new clothes. Or worse. Stephen Chan (2011) offers an especially strong critique based on his own research and diplomatic experiences in Africa. He calls new wars thinking irresponsible, racist, and immoral – primarily because it presents Africans, Asians, and subjects of the Soviet empire as barbarian retrogrades who cater to irrationality in the extreme and who are at home with abominable violence, incivility, and criminality.[4] That argument recalls the types of critiques made of Samuel Huntington's (1993) clashing West and the Rest perspective on international relations, as well as some aspects of conventional development theorizing (Sylvester, 2006). But Chan is also concerned that Kaldor assigns rationality to the old wars of the West and irrationality to wars of our time that are directed by other forces (as though anyone could really argue that World War II applied atomic and holocaustal force in a logical manner). In fact, Chan argues, analysis suggests that each war displays its own forms of rationality, and the question is whether Western scholars can understand those other rationalities or not – whether they even try to understand them.

That is a controversial point, but it is upheld to some degree in research that Maria Eriksson Baaz and Maria Stern (2008; 2009; 2010) have conducted on why soldiers rape during war in the Congo; oddly, painfully, and infuriatingly for both researchers and readers to apprehend, the soldiers have their reasons and can articulate them. While some critics hold forth on the need to treat each new war in its complexity (e.g., Duffield, 2001, 2007; Newman, 2004), rather than lump all into one master category, Chan agrees but also suggests that scholars investigate the logics informing each contemporary war. One might find that the dominant logic in each case is not new at all: so-called new wars might be regrettably familiar in the degrees of "barbarism" displayed rather than socially and technologically retrograde. A final key point Chan makes is that Kaldor's vision of a less conflictual future is very old, very hackneyed, and ameliorative rather than fundamental in the changes it pictures.

Still, Kaldor's interest in the social relations of wars of our time holds promise from the point of view of studying war as experience. Tarak Barkawi (2010) offers a similar admonition more recently: IR should study the social relations of war. He argues that "the discipline supposedly most concerned with the question of war and peace – IR – does not actually study the social and political relations that comprise war and its effects" (p. 7). Rather, IR has yielded too much ground to security studies, ignoring what he sees as strong historical evidence that war shapes social relations within and beyond any specific war zone, wartime, and sector of society – even in times of peace. Heonik Kwon's (2010; 2011) research on experiences of the Cold War in Vietnam that linger and shape social relations in present-day Vietnam echoes that sentiment from an anthropological point of view. He relates stories of families that had sons fighting and dying on both sides of the American war. From the perspective of the Vietnamese state today, only the sons who died fighting for the north are war heroes, and only the memories of those soldiers are allowed by law to be commemorated. One matriarch decides, however, that she is going to reconcile her two dead sons by giving their photographs equal status on a family altar, despite the illegality of that act. She has, in effect, had enough social disruption from that war and sees no reason to render one son unspeakable. Her way of explaining the decision suggests a logic of resistance to what Kwon (2011: 84) calls "the bipolar structure of enmity" that Vietnamese have inherited from the international relations of the 1960s and 1970s. During the war, she says,

> I prayed to the spirits of Marble Mountains that my two boys would not meet. The goddess listened. The boys never met. The goddess carried them away to different directions so they cannot meet. The gracious goddess carried them too far. She took my prayer and was worried. To be absolutely certain that the boys don't meet in this world, the goddess took them to her world, both of them . . . So, here we are. My two children met finally. (P. 84)

Such can be the social relations of wars that do not end when IR says they do. Notwithstanding the penetrating critiques that have been made of new wars thinking, particularly its tendency to describe others as more barbaric and cruel than we have ever been, there are threads to the new wars argument that I believe are worth valorizing and extending. As we will see in the next chapter, the social relations of war is something that Cynthia Enloe has been exploring for years – the idea that wars feature gender relations that can carry on in new and old ways during militarized peacetimes. Hence the name of one of her major books: *The Morning After: Sexual Politics at the End of the Cold War* (1993).

Constructivism and social relations of war: Karin Fierke

Constructivism is not an easy IR approach to summarize – it reminds one of feminist IR in its many instantiations. There are forms of constructivist IR that some call postmodern (Finnemore and Sikkink, 2001) and many would call conventional in their adherence to certain common social science assumptions and methodologies, and there are forms that are postmodern and critical. All constructivist thinking, however, is bound together by the guiding concern that it is important to study aspects of collective social life that are ideational – prevailing norms, knowledge, culture, argumentation and intersubjective ideas in general – if one is to understand international relations. Famously, it maintains that there are agents and there are structures, and these are mutually constituted. If analysts study what these mutualities are, how they are formed, and how they change, it becomes possible to understand or explain "why the political world is so and not otherwise" (p. 393). Constructivism is also seen as answering back to the more materialist traditions of IR analysis associated in the USA with realism and in Europe with Marxism, by arguing that analysts cannot just name and assume the interests of states or other political groupings, national or otherwise. Put differently, one cannot "read" those interests off their structures or material components and locations. Ideas and norms come into circulation in interaction with a variety of particularistic social forces, and these must be investigated.

One would assume that the ideas circulating around, and mutually constituting the structures of war historically and in our time, would be of central concern to IR constructivism. But that has not been the case. Despite its 20-year existence as a self-identified approach to IR research, constructivism has not moved very actively into the study of war. A recent set of exchanges over the online site called political science job rumors (http://www.poliscijobrumors.com/topic.php?id=36426) illuminates that problematic. A writer named anon starts: "I'm looking for some literature that discusses how constructivism explains or predicts war. Thanks." What follows – once any conversation starts at all – is a labored effort to identify constructivist work on war, with possibilities offered ranging across scholars usually associated with realism to various security studies scholars, and tangential arguments about who is a real constructivist and who is something else. One contributor called Techne presents a long bibliography of constructivists that includes just about everyone of note in the field of IR except, oddly, those doing work in feminist IR. As the "conversation" carries on, it gets heated and even a bit nasty as it veers off topic now and then. At the end of the discussion, though, it is difficult to know where constructivism stands with respect to studying war. A contributor called DrRight puts it this way: "The sad truth is, mainstream constructivists have very little of substance to say about war . . . if you're looking for a generic 'Here's what constructivists think about war' paper to work into, say, a syllabus on international security, you're not going to find it. Yet another glaring weakness of the paradigm . . ."

That is most probably a North American point of view on constructivism and war, and an incomplete one given that Martha Finnemore's (2003) work on norms of humanitarian intervention touches on issues of war. Finnemore and Katherine Sikkink (2001: 404) do suggest that there has been bias in constructivist research toward studying the nice norms of international relations, like human rights: "Constructivists in IR have tended not to investigate the construction of xenophobic and violent nationalism …." Across Europe, one finds that the constructivist approaches to security issues can also resonate with the nice norms foregrounded in peace research (Buzan and Lene Hansen, 2009: 191). Certain writings by Karin Fierke, however, at St Andrews University in Scotland, use constructivist ideas to explore social aspects of war.

In a piece on trauma and war that crosses usual levels of analysis, Fierke (2004) addresses experiences of war that leave the ordinary people involved without a means of expressing what they and others in the community have been through, verbally and emotionally. She is especially interested in the kinds of activities that can result in war-related trauma as a widespread social experience (Hiroshima and Nagasaki, 9/11, the Nazi extermination camps). Typically, instead of giving rise to community grieving and mutual compassion, trauma brings on numbing and a solipsism borne of the effort to reduce individual pain and vulnerability by isolating oneself from others. Fierke (p. 479) also notes Wittgenstein's characterization of trauma as an experience "whereof we cannot speak." If one is unable to speak of something, community cannot be (re)established and its history recorded. Fierke argues that social trauma of war can spill over into a political form of solipsism, which she thinks of as a separate level of analysis from social trauma. She explains: "While war involves physical, psychological and political trauma, these are all byproducts of a political context. The three experiences may intermingle as part of a 'politics of trauma,' while remaining separate levels of experience or treatment" (p. 482). It works this way. Individuals grieving the loss of loved ones and the destruction of their environs in wartime suffer individual emotions that are related to the political world that brought them on. When war losses are compounded by a widespread sense of political betrayal, humiliation, or defeat, then grieving becomes a larger social trauma.

Using the example of World War I outcomes leading to World War II, Fierke argues that a defeated nation, Germany, was humiliated and sought to reenact the war and do it properly a second time. By contrast, France was on the winning side and wanted to move away from the war to a more promising future. Fierke reminds us that "a myth developed after the war that Germany hadn't been defeated but 'betrayed,' stabbed in the back by traitors, pacifists, Jews, and, in particular, the politicians of the Weimar government" (p. 485). The Nazi leadership was then successful in articulating the unexpressed emotional trauma felt across the nation. Solipsistic Germany would be born then as a vigilant and separate place to protect from internal traitors and an untrustworthy world. The story does not end there. Giorgio Agamben (1998) argues that exceptions to usual laws on killing can become the rule. In this case, Germany's sense of itself as a special but betrayed and belittled country was passed on to some of its victims. Israel claims an exceptional history that entitles it to be vigilant toward its neighbors and militaristic in its responses to transgression. Fierke's important point is that social suffering does not end with the ending of a war, a realization that Kwon (2010; 2011) also presents in the long aftermath of war in Vietnamese communities today. He does find local people deciding to reconcile the painful past with their needs now, and a similar reconciliation is showcased in recent memoirs by grandchildren of infamous Nazi criminals (e.g., Senfft, 2008), who talk about the silences and emotional chilliness of their upbringings. Unmourned experiences of war, however, do not heal easily.

Fierke's IR argument is that war traumas not only beget individual solipsistic responses but also can afflict entire nations, with possibly dire consequences. Large-scale social solipsism can lead to attempts to "replay the past in order to do it differently, to reconstitute identity and agency . . ." (Fierke, 2004: 188) as a life-or-death matter – as the US government showed in its warrior articulations of national trauma in the wake of the September 11 attacks on the World Trade Center. The value of this kind of constructivist IR analysis is, once again, that it places emphasis on understanding the social forces that can operate in international relations to affect (or presumably bypass) state foreign policies. It works within a basically conventional IR concern to understand how cycles of interstate political violence and war can flow from sources that IR of the realist tradition would consider interesting, perhaps, but not salient. This approach does suggest, however, that the sense of new wars as stamped with extraordinary and new intensities of cruelty, is, in fact, not new to modern wars.

It is a pity that constructivist or other IR research on war as a step sequence of experience that involves people and their states is so sparse. Much conventional constructivist work does not resist the general trend in IR to downplay the study of war experience, even as it usefully directs attention away from realist claims that macro-dynamics like anarchy and abstract rationality can enable the field to explain conflict and war as well as the constraints on war. The particular importance of Fierke's analysis for studying war from peoples' experiences is that it carefully links war trauma experiences that are felt at the individual emotional level to the development of widespread national trauma, to the formulation of militaristic national security policy and then the use of war or other violent hostilities to settle old scores and somehow turn the formerly vanquished into the victors in a new order.

The critical war traditions of IR: Tarak Barkawi/Shane Brighton and James Der Derian

The critical traditions of IR scholarship are no easier to nail down than the constructivist streams, mainly because the two conjoin around their interest in social aspects of international relations. There are several points of departure between conventional and critical constructivism, however, and one is over the role of the state and the significance of rationality to "correct" state or human behavior. The critical tradition looks at processes of meaning-making that percolate through societies to states, states to societies and into the international.

It questions the existence of objective conditions such as rationality or national interest and starts from the alternative assumption that the things we think of as objective are actually power-laden social constructs that often emerge through culturally and historically specific language use. In a critical analytic study of the Cuban missile crisis of the 1960s, for example, Jutta Weldes (1999) compares the language US policymakers used to characterize the events they observed with how the Soviets and Cubans linguistically constructed the same events. That the discourses were different suggests that there would not be one correct version of the Cuban missile crisis, a point that even Graham Allison (1969) raised in the face of inadequate data on the procedures brought to bear on US decision-making at the time. He constructed three possible scenarios, only one of which highlighted the rational state decision-making common in the IR literature. Allison, however, did stay within a social science methodological framework that precludes the idea that international phenomena might be undecidable at all times owing to always imperfect information and fallacies of rationality. From a critical IR perspective years later, Sungju Park-Kang (2011) calls Allison's three scenarios of decision-making "fictions," on the way to making the larger point that fiction of various types has had a place in conventional IR. His is a line of thinking that places discourse front and center as interpretable "data" that can never be finished and ultimately true.[5]

The critical tradition, however, is multiple. Critical constructivists often lean on formal or logical frameworks of analysis when using discourse analysis. Poststructuralists and other post-positivists, of whom Weldes is one, prefer the less formulaic language analysis of literary, historical, and/or genealogical approaches associated with Michel Foucault. But there is much one could say about the larger critical traditions of IR, which include Marxian, feminist, and art and politics developments in the field. With feminist IR perspectives on war treated in the next chapter, this short section spotlights the war-related thinking of three leading critical IR scholars. One is a postcolonial analyst (Tarak Barkawi), one is a student of political philosophy (Shane Brighton) and the third is poststructuralist James Der Derian. Each takes up the old IR topic of war from an entirely new point of view for the field.

Tarak Barkawi and Shane Brighton

Barkawi's perspective steps away from the main IR traditions to focus on how the international and its many relations, including war, look from vantage points in the Global South, as former colonies are called today. In keeping with postcolonial thinking that inspires his work, he sees IR as a social science that has built its knowledge base on Western historical experiences with war and peace. When IR talks about the international system, for example, its reference point is the Westphalian Great Power system of states. Rationality is also a Western ideal that rests on quarantining the distractions of body and society from some pristine mind, with the state an extension of the rational mind operating in an anarchic and dangerous world. Barkawi questions whether people who have lived in the colonies of the West share those assumptions, either culturally or theoretically. It is more likely that former colonies experience international relations as rule-governed in the extreme, no less today than in the past. The so-called rationalities of the colonial state can seem irrational and cruel when studied up from the former colonies. War can also look different. Barkawi (2004a) writes that the study of war in IR has elided with European history and its world wars to the point where wars waged elsewhere are thought of as "small wars" – small when relatively few Europeans are involved in fighting them, even though the consequences of some such wars can stretch far beyond their borders and immediate time periods (as in the Cuban revolution of the 1950s).

In today's world, Western countries have ongoing or enhanced self-assigned entitlements. They can now legally intervene militarily in any country under the terms of the UN Responsibility to Protect Resolution, as a last resort to protect people from harm done to them by their own regimes or by groups of citizens arrayed against each other. Yet in the contemporary wars on terror, which follow a different and older tradition of national self-interest, the military process of ridding Iraq of Saddam Hussein or Afghanistan of the Taliban has been openly accompanied by Western neoconservative fervor to replace autocrats with democratic regimes; these will presumably operate rationally and not by the dictates of religious texts or kleptocratic local leaders. The old Great Powers, convinced of the rightness of their actions in postcolonial states today as in the past, are perplexed when their variously justified interventions are resisted and when the former colonial world actually bites back.

Yet parts of the world made small in IR analyses have been ensnaring former colonial and imperial countries in wars that the West audaciously starts – perhaps to ferret out an Al Qaeda leader – and then morph into standing wars against new enemies (the Taliban) and old (Saddam Hussein). The resistance the Western coalitions face mixes revenge with determined opposition to longstanding patterns of Western control, meddling, and belittling, and the cultural encroachments and insistences of globalization. While the Western coalitions against terror might imagine local people flocking to the democratic model, even as Western shock and awe war tactics are pummeling them, the actual war scene is complex and messy, with many forces, groups, and authority centers elbowing into the picture. Barkawi finds it remarkable that Western leaders cannot fully register these complexities and the contempt that radiates out from such settings for Western states that broadcast belief in freedom while acting militarily in ways that can foreclose local initiatives and cause countless deaths.

Yet he also suggests that these wars, again like ones that existed in the past, can be thought of not as sharp and violent breakdowns in political and economic relations between belligerent states, but rather as forms of connection between groups designated as enemies. Stated differently, wars are not indicative of ruptured interstate balances of power, norms, or structures of conflict resolution as much as they indicate an intensification of social relations. In the era of globalization, interconnections of all kinds proliferate and sharpen, and inequalities become visible to more people. What were "small wars" at the start can intensify through myriad economic, military, and cultural interactions that both precede and emerge with the current war. Indeed, Barkawi (2004b: 156) depicts "globalization as the worldwide social terrain of contemporary conflict." If it is remarkable that the West does not register the contempt that people in former colonies can work up over Western interventionism, he finds it equally remarkable that students of globalization do not write much about wars today as integral aspects of the globalization dynamic. His guess is that the timing of globalization discourse has something to do with this oversight, as the term came into fashion following the Soviet demise, when analysts were anticipating a peace dividend and positive linkages from greater trade, travel, common language use, and the diffusion of Western culture (Barkawi, 2006, 2004b). Like Kaldor, Barkawi thinks of globalization as primarily benefiting the few against the many, which from his point of view makes war more likely and more intensely interconnecting.

Barkawi (2011: 16) also takes IR overall to task for shifting so much of its attention to security and away from the direct study of war. War studies, he says, should be the branch of IR that foregrounds "politics, force, and war." Writing with Shane Brighton, he adds that it should be critical in orientation and canvas "war, the threat of war, and the preparation for war ... [as] the origins, transformation, and end of polities" (Barkawi and Brighton, 2011: 126). They both argue that because war is a form of social relations, it encompasses gender relations, economies, and technology, and has the capacity to reorder knowledge, social identities, and public reason. Critical analysis is required to evade the search for theoretical certainties in an ensemble of connecting social relations. Brighton (2011: 102) especially emphasizes that practices of war can reveal more about being than knowing because in war, "familiar or taken-for-granted objects of knowledge and structures of meaning are overwhelmed and transformed." He notes that Emmanuel Levinas (1969) anticipated the disordering aspects of war that remake people's sense of themselves, their identity, and their relationship to what they once thought were normal and reliable institutions of peaceful society. His phenomenological perspective is needed today, for it is clear to Brighton (2011: 103) that to study war critically and with more rounded views, there is a "necessity of descriptive, reflective engagement with experience ..." But what that involves is a question that Brighton has not yet addressed and must be fleshed out.

Although much of the effort involved in understanding war in the ways that Barkawi and Brighton suggest entails going beyond IR into knowledge generated in other fields, the two writers are not keen on interdisciplinary approaches to war studies per se. They argue that existing war studies courses or writings are chasing a subject and set of research questions that its component fields have not been able to pin down on their own. Indeed, the group of interdisciplinary scholars associated with a project that I facilitate called Experiencing War (Sylvester, 2011a), express a sense that they feel "more or less isolated in their parent disciplines," a sense that Barkawi and Brighton (2011: 131) have also found, and also feel very isolated within

IR, including its critical and even some of its feminist arenas. Barkawi looks forward to a new field of war studies within IR that can dialogue with security studies rather than isolate the one from the other. In many ways, dialogue is already happening in the war studies sector of feminist IR, as we shall see in the next chapter.

James Der Derian

James Der Derian might be one scholar of IR who carries on studying war with particular panache and produces work that is novel and unmistakably his. Der Derian's best-known study is *Virtuous War: Mapping the Military-Industrial Media-Entertainment Network* (2009), which takes a hard critical look at a different network of war-making than is discussed in new wars theories. That network of academics, media, and military did not challenge the Bush administration's ill-advised decision to invade Iraq or decry its empty assertions that Saddam Hussein had weapons of mass destruction. Although much of the book has the ring of strategic studies, which strays very little from analyzing policies, technologies, and games of war, one of its most innovative aspects is its grounding in Der Derian's family experiences with war and his own personal/professional efforts to talk critically about the impending Iraq War at a time when everyone in the network seemed hot for it. It is a work that builds bridges between the tough emphases on state-centric power politics that have always been associated with IR and the out-and-out critical traditions that have little time for any of the old IR. The unexpected thread tying these together is story-telling as a quasi-anthropological methodology that telescopes very physical, psychological, and social aspects of war.

Starting with virtuous war stories featuring his grandfathers, Der Derian turns sharply into the present time of war, with its virtualities flying high. It is an era where software prevails over old hardwares of war, and new hardwares of war produce the robots that P.W. Singer (2009) has recently detailed in accessible ways and the drones that often replace flesh-and-blood soldiers to deliver precisely targeted injuries to people and to infrastructures. Add the public, private, business, academic, media and information agents of war that he calls MIME-NET. All of this new networked and virtual capacity is employed for the virtuous goal of spreading democracy and markets to places that Barkawi insists are resentful of such impositions. Der Derian calls it "ethical change through technological and martial means ... [or] the technical capability and ethical imperative to threaten and, if necessary actualize violence from a distance – *with no or minimal casualties* (p. xxxi, emphasis in original). Technologically, he says, we can now kill "over there" without "worrying" about our responsibilities in killing – the deaths are by pilotless drones, not by bombs dropped from planes we are flying; or, as in the case of the raid on Osama bin Laden's house in Pakistan, the mission can be conducted with such precision and individual anonymity that the SEALS team involved might just as easily have been drone-clones.

Der Derian travels the terrain of virtuality in war by personally following American military exercises into the Mojave Desert, talking to officers and business leaders about relevant information technologies, observing computer jocks gaming war in Tampa's Central Command, and learning about "peacegaming" by the military and about the Synthetic Theater of War (STOW) planned "to integrate virtual, live, and constructive simulations of war in real time" (p. xxxvii). He goes on to the University of Southern California where an Institute for Creative Technologies brings together military, computer, and entertainment industry executives, gamers and graphic artists to think about the future of war. All this war activity is emblematic of the postmodern sense that truth and certainty are slippery, especially in the hands of the state, and that it is as important to study war up from the ground allotted to militaries as it is to study down from the isolation of a military computer room or a professor's university office to people on that marked ground. Although his research takes him to the hard places of realist practice, a tradition he does not entirely disrespect, Der Derian ultimately describes his thinking about studying war in terms that could not make sense to IR theorists of realism:

> this book is as much a detective story as a cautionary tale. Scholars and journalists have been slow to cover the story of virtuous war, mainly because they can't find the smoking gun, let alone the increasingly virtualized body. I don't know where the bodies are, and, from my own family history, know too well the significance of when they go missing. I committed myself to wander in deserts real and virtual because I believe most profoundly, as Walter Benjamin did in the waning days of the Weimar Republic, that "in times of terror, when everyone is something of a conspirator, everybody will be in a situation where he has to play detective." (P. xxxviii)

"Everyone." Remarkably for someone in IR, Der Derian is including bodies, families, and presumably a range of other people who could be involved in virtuous war and suggesting that we all must operate with the agency and methods of detectives and those who write about them. That is a very generous attribution of significance to those that IR usually ignores – except that Der Derian seems to be operating in his book in a world plastered with military men.

Der Derian uses a lot of images in his writings, some created out of his unique way with words and some from the lenses of cameras, his own and others. The reader might be looking at pictures of his Armenian ancestors and then come upon Nancy Burston's photograph, 'Warhead 1,' which he describes as "a digitised composite of world leaders proportioned according to their country's nuclear weapons, in which the facial features of Reagan (55% of the world's throw-weight) and Brezhnev (45%) dominate the fuzzier visages of Thatcher, Mitterrand, and Deng (less than 1% each)" (Der Derian, 2005: 23). Or, he presents a word image of the contemporary terrorist as someone who can now easily do double-duty as an airport security profile, "featuring the checkered keffiyeh of Arafat, the aquiline nose of Osama Bin Laden, the hollowed face of John Walker Lindh, the maniacal grin of Saddam Hussein, the piercing eyes of Abu Musab Zarqawi . . ." (p. 27). After a career history of intermingling such devices with cartoons, drawings, and his own diaries, Der Derian is now co-writer and co-producer of a documentary film that shows aspects of MIME-NET in action. The film is *Human Terrain* (Der Derian, Udris and Udris, 2009), and is about the US Pentagon plan, implemented with $4 million of government money, to set up a Human Terrain System consisting of academic cultural anthropologists embedding with soldiers on the front lines in Afghanistan and Iraq (and on the desks behind lines). The idea was to win hearts and minds through providing military personnel on the ground with a better understanding of local society and culture. The film features fascinating scenes of military men expounding their theories of the social relations of war while the anthropologists work to teach those men about the cultures within their war zones (and one of them dies in the process of doing so). If MIME-NET impresses as a million sophisticated networks networking each other, the Human Terrain System is cringe-making, the kind of do-good naivety that gets the USA into wars for democracy or regime change – with the resulting social relations of contempt-cum-interconnection that Barkawi outlines.

For analysts who seek ways to approach war that do not reduce a complex phenomenon to state militaries, weapons and the like, Der Derian's work is highly instructive. For, while professing to find most of IR annoying if not risible (he describes constructivism, for example as something that could have come from outer space (2000: 782)), he is noteworthy in part for remaining an "IR person" come what may, for moving back and forth between and among traditions that he cannot fully embrace, and for fearlessly – fearlessly – insisting that the relations in international relations are both social and technological, state-centric and people-centric, spectacularly real and spectacularly sci-fi all at once. His work is cocky, but brilliant in partnering so many seemingly opposed groups in MIME-Net, relying on primary research gathered from US military officials, while also including stories and photographs of his Armenian grandpa. Next time, though, show us grandma, too.

What emerges?

This is a very small sample of the IR literature on war. We can begin to see in and across it, however, a range of concerns that could fit with the orientations of this book. There are hints of a more social side of war embedded in the structuralist emphases of contemporary realism. There is some evidence that constructivist approaches can foreground people's experiences with historical war (e.g., trauma) as having agentic qualities that map onto future conflicts (Fierke). It is also noteworthy that the critical IR camp can write very directly about war as a socially networked institution (Der Derian) that connects battlefields with wider social fields in ways IR urgently needs to consider (Barkawi and Brighton). The next chapter will detail work in the feminist IR camp, also in the critical tradition, that develops particularistic approaches to war grounded in experience(s) of people, especially women people.

Curiously, no one in this short line-up of IR raises directly the point that war is a social institution. Barkawi comes the closest by advocating that IR engage the social relations of war. Der Derian's elaborate MIME-Net could be seen as an aspect of today's social institution of war, except that it leaves gender relations, and the Global South, more or less out. Stepping around the notion of war as a social institution also characterizes much of mainstream contemporary sociology, something that Sinisa Malesevic, in his *The Sociology of War and Violence* (2010), finds equally odd. He laments that "not only are there no established specialized journals or professional organisations within sociology that focus exclusively on warfare but there are very few, if any, books and journal articles that study the relationship between social structure, agency and wars or other form of organized violence" (p. 50). Of particular relevance here is Malesevic's sense of the implications of this forfeiture beyond sociology: "this neglect within the discipline has created a situation where an overwhelming number of studies dealing with warfare and organized violence lack any sociological grounding" (p. 50). His characterization of mainstream sociology does not carry over to its critical wing, though. The journal *International Political Sociology (IPS)*, which is affiliated with the International Studies Association (ISA), the largest organization of scholars of international relations, broadly understood, regularly carries articles on facets of war. Sociologists per se do not necessarily write these, but most articles engage sociological concepts of a critical European bent, which means that the key ideas are grounded in contemporary continental philosophy.

The June 2011 issue of *IPS* alone carries three articles with "war" in the title, including the piece by Barkawi and Brighton I have been referencing. The lead article by Michael Shapiro is particularly salient because it raises issues about war, experiences, social relations, and gender that bridge IR traditions of war, with their sly hints of the social, and war studies in feminist IR, which foregrounds experience and the social-gender relations of war that even critical IR work can treat with passing gestures rather than serious study. Shapiro's piece starts with Hannah Arendt's discussions of the split between domestic and public spheres that dominated philosophical thinking until the state started managing economies. The private and public spheres then flowed into one another "like waves in the never-resting stream of the life process itself" (Arendt, 1958: 31 quoted in Shapiro, 2011: 110). Shapiro imbues her ideas with a revised directionality, arguing that aspects of contemporary war politics, in fact, can start within, and flow out from, households. He uses the example of Cindy Sheehan's household, which she effectively moved from a private space into the public near George W. Bush's estate in Crawford, Texas, and then made her action into a double private–public performance by purchasing private land near the President's estate and carrying on her public anti-war protests from there.[6]

This is an unusual starting-point for an article that is not treating an element of gender as its main topic. What intrigues Shapiro in the Arendtian claim, which resonates, of course, with a variety of feminists' stances ranging from Marxist to women in development thinking, is the idea that to be in politics one must move out of the private sphere and into the public sphere. Sheehan's activities belie that claim and fall more into line, in Shapiro's view, with Jacques Ranciere's perspective that politics does not depend on where anyone is located spatially. Politics only comes into existence when some action is taken to correct some injustice. Politics is an event involving dissensus, disagreement. Political events, understood in that way, "reorder spaces and reconstitute identities, rendering persons as political subjects" (Shapiro, 2011: 112). Shapiro is interested in how the Sheehan case illustrates the way some households have become sites of a micropolitics that registers war as "space–body relationships" (p. 112).

Bingo. Shapiro is exploring intersecting relations of personal, social, gender and political experiences (events of dissensus) of war, taking his cues openly from a woman's experience with a social institution of war that has come into her space and she into its space as politics. The questions he goes on to ask are nearly the same that Barkawi and Brighton raise about studying war critically as social relations, are inflected with Der Derian's sense of networks urging war, and are nearly the same as the queries that guide the *Experiencing War* project and book (Sylvester, 2011a), and the introductory concerns expressed in this book. He says:

> As the US wars proceed in Afghanistan (and until recently Iraq), their impacts are experienced very differently in the diverse lifeworlds of the participant/ observers. Some are at the war front, exposed to the actual "theaters" of war, where all in close proximity are in mortal danger; some are at home fronts, attending cinematic theaters, where their participation involves consuming images and narratives from a safe distance; and some experience the wars vicariously, playing video games that interpolate them as warriors with the same level of actuality as those "service personnel" who operate lethal weapons from computer consoles at a great distance from the war theater. (Shapiro, 2011a: 112)

My concern with rethinking war, however, is not limited to wars that the USA conducts; and presumably Barkawi and Brighton, who come from a postcolonial and globalization perspective, are not either. But all of us are interested in variously situated war participants and observers, as well as their experiences as part of what Shapiro calls the presence of war. Aspects of sociologically-inspired thinking thereby seems to lay an emergent basis for the comeback topic of war in and around IR – on new terms, to be sure, but hopefully with input from older traditions that cannot help but slip people and social concerns into their more abstract analyses in order to strengthen (by humanizing?) their points. And crucially, there is strong input from feminist IR, as it is noteworthy that even the sociologically tinged work on war manages to overlook all the new feminist IR war studies research.

"Bingo," therefore, is accompanied by a mind-boggle: how could feminist literature be so invisible to those who want to study aspects of households, social relations, and experiences of war? Possibly the answer is that the social science disciplines interested in war have been training its students, over and over, to have an aversion to people as agents of and experiential participants in the international and its relations – such an aversion that when I spot people at all in IR war research, let alone women,

I get excited.[7] Most are disinclined, as the anthropologist, Veena Das (2007), puts it, to descend into the ordinary for (gender) insights into the extraordinary. I am reminded at this juncture of an opinion piece the *New York Times* columnist, Joseph Hallinan (2011: 10), wrote around the question of how bright financial experts could possibly fail to register cascading mistakes that would bring on the Wall Street crisis of 2008. He says that the financial experts all misread the "data" the same way and that this type of group error – and here I am extrapolating to the IR context of surprise that events of the Cold War would turn out as they did, and to the blank spots about feminist IR war studies – is "most likely to be discovered by those who . . . look at the world with new, unblinking eyes." We turn to some unblinking eyes next.

2
FEMINIST (IR) TAKES ON WAR

War has many forms as well as numerous official and fugitive centers of authority; and gender relations feature, of course, in all of them. Many people at least tacitly associate war with militaries and militaries with men. Certainly some of the early philosophers, whose names regularly dot IR texts, based their sense of politics on such associations. To their credit, Machiavelli, Hobbes, and Rousseau actually theorized gender relations in their discourses on politics, something contemporary philosophers and IR theorists are less likely to do.[1] That early gender thinking, however, established people called men as the active political humans and people called women as proper to a sphere of private morality kept separate from the man-sphere, or likely to be conquered by it in any event (Pateman, 1988). Thereby, it would be seen as normal in many parts of the world for men to be at the helms of state and directing the politics of militaries and war. Commonplace understandings of war today can still be starkly sex-differentiated: men do war and women suffer, support, or protest war.

And so it can seem. Even when women are admitted to military organizations, they can be kept behind the lines or out of range of collective acts of violence that define war. Military commanders have argued strenuously over the years against recruiting women into Western fighting forces, claiming that their presence will distract and endanger male soldiers, lead to reduced combat fitness among troops, and weaken the military overall. Women's bodies, it is said over and over, are not big enough, strong enough, and emotionally equipped to deal with soldiering. Their supposedly body-based "instincts" also incline them toward protective roles and not aggressive attack modes. A parallel line of gender "reasoning" has been applied to gay men in state militaries: to paraphrase only recently rescinded US policy, "just don't tell anyone." And babies: who will have the babies and protect them from harm if both women and men go off fighting? Indeed, will there be anyone left home to protect through war?

Gender and war is a very fraught coupling, and the men and women who join the institution of war or work with it can put observers in a conundrum: to support them, not support them, take no position, look the other way? There is unease among feminists in particular about khaki and how it can become some women (Enloe, 1983), just as there is unease about women who self-identify as feminists but depart the common feminist advocacy of peaceful conflict resolution. As noted earlier, feminists historically number among the strongest supporters of the commonplace association of women with peace – not necessarily for the biological reasons that classical philosophers foregrounded, but for feminist movement reasons. Feminism, it is also said over and over, is against violence of all sorts. As part of the lingering sense that men are either hardwired for war or, as is more likely the case, socialized to it, feminist analysts have shown considerable interest in understanding celebrations of warrior men and masculinity (e.g. Parpart and Zalewski, 2008; Higate, 2003).[2] Not all feminists doing work in IR today, however, are concerned to study peace rather than war or to take up the masculinities and war emphasis of recent years. Some think that equating women with peace continues a long tradition of ignoring violence by women or subsuming it under a category of gender deviance, as would very likely have been accepted in the days of the classical philosophers. Moreover, a great many women in today's world have no choice in the matter of war or peace. War comes to them in the so-called new wars that target civilians, and war also provides some women with opportunities to pursue political goals through violence.

This chapter draws out the main concerns and arguments articulated in two early feminist IR works on gender and war and then moves to three current feminist IR war studies. The early traditions are represented in work by Jean Bethke Elshtain (1987) and Cynthia Enloe (2000; 2010) and by broader feminist discussions about what experience is and its reliability as a tool of insight. The discussion then shifts to a post-9/11 time-frame, when, after a long lull in feminist writings on war, terrorism thrusts war front and center for some well-known feminist analysts, and a new generation of war studies emerges in feminist IR against the backdrop of ensuing conflicts in Afghanistan, Iraq, Africa, and South Asia. Within the feminist IR group, this chapter focuses on exemplary books by Miranda Alison, Megan MacKenzie, and Annick Wibben. Foregrounded throughout are the methodologies each generation uses and the strong emphasis across feminist traditions on ordinary women and their experiences of war.

Elshtain and Enloe consider war

Jean Elshtain and Cynthia Enloe are two of the original feminists of IR who took up the field's core concern – war – in critical ways. Cynthia also brought IR questions to feminism through her writings and tireless feminist peace activism. Jean has more recently written on just and unjust wars and publicly backed American-led interventionist wars following 9/11; her work has antagonized today's generation of feminist and security studies analysts, resulting in some sharp exchanges (e.g., Burke, 2007; Elshtain, 2009; Sjoberg, 2011). Enloe, a figure of admiration in the same circles, has a do-no-wrong reputation that is rare among politically engaged academics. Their very different approaches to war are prototypes of the war studies emphasis that emerged in feminist IR in the latter years of the twentieth century.

Jean Bethke Elshtain

In *Women and War*, Elshtain (1987) juxtaposes philosophical writings on war with accounts of individual women's experiences with war. Elshtain starts with the Hegelian distinction emphasizing men as just warriors and women as the beautiful souls who are off in the background. Her point in doing so is often misunderstood as indicating that these longstanding binary assignments around war are reasonable. In fact, they are prototypes and straw figures that her chapters disembody and re-embody in more realistic and women-affirming ways. She writes that Western men (the only ones on her radar) plan, conduct, participate in and then narrate war to those who have not experienced it directly. Western women (ditto) are judged by society as too soft, motherly, and disinclined to violence to be effective warriors. They are the ones the men war to protect, a point Judith Stiehm (1984) has long made, and also function as the designated weepers and mourners of war, which Judith Butler (2004b) explores many years later. In Elshtain's research, however, no one is where he or she is supposed to be in war, now or at times past.

Elshtain presents a tapestry of war in which some women are actually or virtually in war and some men remain at home. It is a world in which good (Western) mothers share values with good soldiers, and women are as resourceful in defending themselves as men are in attacking. Her evidence comes from actual wars, representations of wars in fiction and drama, and presentations of gender and war in films. Starting with the Greeks, where much of IR locates its beginnings, Elshtain considers warring Athens and Sparta as well as Homer's and Plato's words on war contained in the *Iliad* and the *Republic*. Women emerge in such sources as far tougher about loss and the importance of war than the commonplace gender distinctions seem to reflect. The reformation of the Middle Ages banishes images of saints from the church and thereby renders women's piety-cum-ferocity (think Jeanne d'Arc) invisible as points of reference in thinking about women and war. Elshtain reminds us of Machiavelli's concern that states could be ruined on account of women, and Rousseau's utter conviction that women were bundles of sweet political incapability. The nineteenth century appears in Hegelian celebrations of war, the violent struggles that Marx and Engels equate with human progress, and Clausewitz's heavy lean on his wife as he writes tracts on war *sans* attribution. Of particular poignancy for the present project is Elshtain's reckoning with US pioneer and civil war mothers and wives, who, left alone as the men advance north or fight in colonial wars, work to resist Indian and union forces. Women in other places and times, such as Europe during World War I, take on similar paramilitary home protection tasks across Greater London and even follow some men into war. American suffragettes come out for that war and women war correspondents go to the front.

Elshtain is respectful of official fighting forces but also tells the reader that "the man, and the woman, 'in the street' often knows how fragile it all is, how vulnerable we all are" (p. 91). And, indeed, she includes herself and her experiences of imagining war, thinking about the Vietnam War, and cramming in the works of IR realists as part of the "in the streets" experience. By placing herself in war, though at a distance from any war front per se, Elshtain boldly illustrates the feminist methodological principle of revealing one's positionality in the research. That is to say, the researcher must explain to the reader what influences s/he brings to the research and how background and pre-formed views could affect the findings. Elshtain also foregrounds herself as every bit a warrior on her own turf and is clearly not of the feminist stream of thinking that opposes war. What she opposes are the gender-guarded stories of war as men's business. She argues forcefully that male violence is channeled through war and women's violence is entirely disallowed: "she" has no social institution that accommodates her violence and so seems out of control when committing acts that would win men a war medal. Yet just as quickly, Elshtain points out the scores of men who refuse war by deserting or claiming conscientious objector status. By turning away from their defined space in the institution of war, they too can seem out of control.

This is where her linking of the good mother with the good soldier enters the text. If we think about these apparently opposite roles in society closely, she says, it is possible to find overlaps: the good mother sacrifices for her children and the good soldier sacrifices for country and comrades; the good mother feels guilty about her parenting skills and the good soldier worries that he might have saved a colleague from death if he had acted sooner; women are excluded from war talk and men are excluded from baby talk. Apparently the good warrior and the beautiful soul actually conjoin – and most ironically so, given the usual commonsense gender separations that abound around war. Elshtain (1988) tops up the argument by writing elsewhere that peace is not a separate realm from war: those realms conjoin too. Of course, one could raise objections to the good mother–good soldier equivalency. For one, it assumes that wars of the past and wars of the present revolve around state militaries and home-front mothers, rather than scores of participants who confound the civilian–military distinction. As a second point, how "just" can warriors be if they rape women as part of a war strategy or for other murky reasons? Third, Elshtain's "women" show no marks of race, class, age, or varied global location. Their common identities boil down to two overlapping categories: mothers-women. As well, if the good soldier bears a strong resemblance to the good mother, and she is simultaneously a good soldier, where are we ontologically with these statuses? And what of good fathers? Is there nothing about them because they are assumed to be – all of them – soldiers, or otherwise absent?

The strength of *Woman and War* is that it confronts the gender shibboleths that put men at the center of the social institution of war and put women in various places of support or protest off-stage and ontologically outside war. It is the first book specifically within the field of IR that does this.[3] If Elshtain failed to foresee that World War II would be the last war where uniformed state militaries face each other on battlefields, so did nearly everyone in IR at the time. The bigger issue in her book, which a new generation of feminist IR thinking on war avoids for the most part, is theorizing from a base in feminist maternal thinking, the kind of thinking that has women's private roles seeping into the public realm in ways that are said to be uniformly progressive. It would only be later in Elshtain's career that deeper conservative inclinations would emerge, including moral realist defense of the US-led wars against terror and concerns about same-sex marriage. Both positions have locked her out of most feminist circles and landed her on the shelf of "difficult" feminists or post-feminists.

Cynthia Enloe

Cynthia Enloe was writing her classic *Bananas, Beaches, and Bases: Making Feminist Sense of International Relations* (1989) during the same period that Elshtain talked about women and war. Not unlike many first-generation feminist IR writers, Enloe worked to reveal the international as multispatial and full of IR-relevant people, and relations that would not fit into the few categories of actors and behaviors that the field then provided. She too has probed war, militaries, and women in order to find out where the women are in war, what it is like for them in state militaries, and what other unnoticed jobs women perform in international relations. Elshtain finds women's heroism where it is not supposed to be in war, but Enloe notes the mostly unheroic things women do to keep daily international relations, and its wars, operating. Enloe says that the work of average women everywhere enables the field of IR to keep its eye on the derring-do of elite men, whose clothes are washed for them, beds made, food cooked, documents typed, seating charts prepared, notes taken and so on. It is in realms of unheralded experience of international relations that Enloe begins to develop the concept most associated with her work: militarization.

Her pathway to that concept starts with studies of nationalism, a political phenomenon she sees as having some pluses and many minuses for women. Nationalism can be spurred in part by a local awakening to the way that colonizers treat women; however, "nationalism rarely if ever takes women's experiences as the point of departure, the rallying cry" (p. 44). Likewise, while armed national movements of the twentieth century often promise to promote women's interests during and following an anti-colonial struggle, this rarely happens: some nationalist wars never end (Afghanistan, Palestine); some usher in local governments that become oppressive over time (Zimbabwe); and others carry forward abusive gender practices from colonial and earlier times (India, Pakistan) or start up new "traditions" for women (Taliban, Afghanistan). Even when women serve in forces fighting colonialism, men-and-militaries monopolize the new political contexts and often reinforce social institutions they ostensibly fought to change.[4] With the break-up of the Eastern bloc in the early 1990s, IR became fascinated with the new era of globalization that replaced the strictures of the bipolar Cold War. But along with increased opportunities for trade, travel, and communications came the political economy of war waged with leftover Cold War and national liberation weapons, which could be readily replenished through arms fairs.[5] Enloe's view of globalization as socially connective of enemies and friends resonates with Tarak Barkawi's and Shane Brighton's ideas from postcolonial and critical IR.

Militarization, however, rests less on nationalism, arms and wars per se than on molding people to accept the use of force in the preservation or overthrow of groups, regimes and nations. Enloe (2000: 3) writes of it as "a step-by-step process by which a person or a thing gradually comes to be controlled by the military or comes to depend for its well-being on militaristic ideas." Men, she argues, are not "natural" soldiers. They must be socialized into believing that military activities are manly or honorable, and that military solutions to intractable political disagreements make sense. Men must be honored as heroes when they fall in wars and seen as exemplary citizens when they survive war. Enloe calls this phenomenon militarized masculinity. The militarization of women occurs more often through global popular culture outlets, like violent Hollywood films that feature men with the guns (for the most part) and women who are attracted to that fierce masculinity. In the West, she might happily ride in cars that look like armored personnel carriers and don clothes that mimic – and tout – a military look. She will still be paid less than men in most work situations and yet be expected to support defense spending. The quirk in all this is that militarization becomes acute at the *end* of the Cold War and before the global war on terror begins, a ten-year period that the IR realist-neoconservative Francis Fukuyama (1989) celebrates as the beginning of the end of epochal ideological conflict and war. That is the time when militarization accelerates, to the point where security measures at airports, for instance, grow increasingly intrusive and unwieldy, but are increasingly accepted, too. Enloe's writings alert readers to the power required to make militarization normal, an orientation that even realists in IR uniformly ignore.

Without saying so directly, the militarization that Enloe talks about is part of a pervasive social institution. It eats up resources, invades all news, feeds a voracious and multi-located arms industry, and too easily comes to mind as best practice for handling political disagreements and intransigent regimes in the Middle East, South Asia, Africa, and parts of South America. The wars of our time are no longer nationalist per se; they involve various coalitions and networks of local and international forces tied to (and sustaining) global political economies. Recall that Vivienne Jabri (2010: xi) names war as the chief international political matrix of our time. In her words, it is "a set of interconnected practices – these might include war, invasion, incarceration and deportation, the use of torture, the surveillance of individuals and communities as these move both within and across state borders, and pedagogical exercises aimed at state-building – involving states as agents, international governmental and nongovernmental institutions, quasi-official and private, engaged in a complex transnational network recruited in the service of a global machine the aim of which is the government of populations." Jabri enumerates the many places that Enloe can sight militarization, and the gender relations that tether men and women everywhere to violent orders.

Enloe (2010) takes that theme forward in her recent discussion of the Iraq war from the perspective of four Iraqi women and four American women, each of whose circumstance becomes the basis of a larger discussion of some aspect of that war. Written in Enloe's signature style for a wider than academic readership, the book points out two new lessons about war that she has learned from following these women and their families, friends and situations through accounts of them in the media. One is that all wars take place at particular moments in national and international gender histories, and those histories set the parameters for the women in them. The second is that each war is characterized by phases of gender politics, which shift the balance of opportunities and dangers for men and women. Enloe provides a poignant example of both forces impinging on one Iraqi woman's life:

> For instance, Sabriyah Hilal Abadi had been a seamstress before the war. Her husband had worked in a government-owned factory. Together, the couple had brought home enough income to support their four children and pay the rent on their apartment. Then Abadi's husband became a victim of the kidnappings that sowed fear in urban neighborhoods. In July 2005, after armed men pushed her husband into a car, she never saw him again. Now husbandless, she had been forced out of her apartment. Being resourceful, Abadi had taken shelter with her children in an empty, dilapidated Baathist Party building in the Zayouna neighborhood of Baghdad. Soon other newly homeless women and their children joined her. But then the new government's soldiers forced her out, claiming that Abadi and the other women and their children were squatters, that this was a government building, and that squatters like her were the cause of crime in the area. At that point, Abadi accepted an AK-47 rifle from friends who were worried about her security. Before the war, she told a reporter, she never would have held a rifle. (Pp. 66–67)

Enloe is describing a process by which an ordinary woman becomes party to militarization around her. She is also doing some of the descriptive and reflective work of engagement with war as experience that Brighton urges.

There are some quirks, however, in Enloe's writings on war. One is her steadfast position against war, which I hasten to say is an admirable advocacy position, but one that has the logical consequence of putting women who embrace war, or use it to pursue political goals, in a difficult analytical and experiential position. If militarization is the temper of our time, and militarization is everywhere encouraged and is always a pernicious politics, then women who freely enter war and its apparatuses must be suffering a latter-day version of false consciousness. They must be buying into the matrix of war, or be oblivious to its ubiquity and dangers, or believe that war and militarization benefit them or their societies. Militarization thereby becomes bigger than life – a metanarrative like capitalism that permeates everything and accounts for more than any concept can. Alternative practices would not likely survive the onslaught of messages, production schemes, and socialization processes that spin around this metanarrative. Militarization becomes an abiding global rule; and that can seem both reductionist and essentialist. Enloe's recent book tempers these views in recognition of the many ways women become involved in and experience the Iraq War. Her argument there resonates strongly with the concerns that Butler (2004b) expresses in *Precarious Life* about whose stories count in war and whose are marginalized, and how war keeps women on both sides of a war from discovering connections with one another. Butler, however, is not mentioned in Enloe's book.

Enloe's analysis demonstrates the value of post-structuralist emphasis on problematizing modes of social and political discourse that seem normal and defensible but are terribly flawed. War and militarization, Enloe insists, are not normal. Both are created and interrelated in ways that require considerable effort to sustain. Effort will also be required to demilitarize our minds and practices and delink national security, vulnerability, militaries and militarized masculinities. Yet Enloe's emphasis on a totalizing problematic of international relations – militarization – frames the discussion as always already obvious to those who have feminist curiosity (Enloe, 2004). Can one be a feminist and yet not buy her militarization/demilitarization argument entirely? Or could that be a difference that is not given voice? There has been no real debate on these war questions in feminist IR, which means that the experiences of many women in war have not been taken into account theoretically and empirically, to say nothing of feminists who might want to study war and not lead with militarization or militarized masculinity. The politics of women who might be curious in unapproved ways can be sidelined in feminist analysis and that can be both unjust and an obstacle to nuanced thinking about collective violence. Enloe's inspiring work, and Elshtain's latter pro-war conservative views, brings to the fore a key tension in contemporary feminist theorizing: how to take a political position and also incorporate positions that differ from "ours?" Whose experiences of war and militarization are the ones feminists in IR should embrace?

Whose experiences? What experiences?

The case of Aayan Hirsi Ali is fascinating in this regard. To me, her experience is a touchstone for questions about whose views count and which experiences (and curiosities) feminists take on board today relative to those that flunk various litmus tests. Ali is the Somalian-Dutch feminist and former MP of the Netherlands who writes antagonistically about the effects of Islam on women adherents. Her personal story is remarkable. An immigrant to the Netherlands, she escaped a marriage plan devised by her father by leaving her flight from Somalia to Canada when it refueled in Germany and making her way to the Netherlands as an asylum seeker. She learned Dutch, went to Leiden University and, still in her thirties, became a Dutch member of parliament. If one is seeking a feminist heroine, surely this chain of events qualifies Ali, especially as she is a declared feminist. But no, Ali is best known for speaking about the burdens Muslim women bear in their religious cultures (Ali, 2006, 2007, 2010) and for being punished in the Netherlands for doing so.

Ali endorses the late feminist theorist Susan Okin's (1999) critique of Western policies that seek to protect the integrity of besieged cultural groups over individuals in them. And she has not been gentle in her critiques. Ali worked with the Dutch film producer Theo van Gogh to produce a short film called *Submission I*, in which violent, anti-women sentiments from the Koran appear on the backs of naked women. Dutch society became alarmed as Muslims in the Netherlands from predominantly Turkish and Moroccan backgrounds condemned her. One among the latter assassinated van Gogh as he rode his bicycle to work on the streets of Amsterdam, thrusting a knife in his chest with a message of warning on it to Hirsi Ali. Almost overnight, Ali became persona non grata in the Netherlands. Indeed, the degree to which one African woman could insecure a "tolerant" Dutch state grappling with significant Muslim immigration is remarkable. Ali's application for citizenship, successful more than a decade earlier, suddenly came under state scrutiny for possible errors. In 2006, she was stripped of her MP position and Dutch citizenship on the grounds that she gave a false surname on her immigration documents. Ali had openly spoken of her decision to use an extended family name on the application rather than her own surname, so her father could not find her and pursue the arranged marriage. To do so is not against Dutch law, and a shamed Dutch government later reinstated her citizenship. Ali now lives mostly in the USA, where she is a fellow at the American Enterprise Institute, an organization that US feminists tend to disdain.

Ali believes that she is taking "an enormous risk by ... joining in the public debate that has been taking place in the West since 9/11. ... I must say it in *my* way only and have *my* criticism" (Ali, 2006: xviii). It is a risk that has not earned her the support of many Western feminists, who worry that she upsets women in Muslim communities, tries to tell those women what to think, and works now for a conservative think-tank in the USA. That Ali is from the religion she critiques but lives as a difference to it, that she encourages public debate about religion and Muslim women's human rights, seems less important to feminist critics than her personal style (harsh) and her deviant position within a group that is haunted by events of September 11, 2001. Dutch feminists have called her an ambitious woman manipulated by the racist right, or a bitter woman whose personal circumstances are the only basis for her condemnation of a world religion. That many sided with Dutch Muslims and the Dutch state against Ali shows a monumental shift in feminist thinking, away from defending besieged, outspoken and religion-questioning feminists to aligning with a nonfeminist cultural majority and mostly male state against feminists who upset them both. Open debate about which positions (in the plural) make sense for feminists living in complicated times can be quickly shut down either by strong views one way or the other or by reluctance to take any position at all in a situation of cultural pluralism. Hirsi Ali embodies contradictions in feminist thinking faced today. She is denounced across wide swaths of Western feminism for forcing difficult and complex issues into the open and for doing so her way, not "our" way. But is that not what difference is about – being open to a variety of experiences (Sylvester, 2010a)?

That question about an individual feminist difference segues into another that highlights issues of experience and difference as they relate to war: what constitutes a recordable, respectably different, and perhaps valorized feminist war experience? Must a woman protest and resist war to fit feminist thinking or can her difference be participation in war, including killing? Feminist research practice is often about experiences – women's experiences, mostly, which are less commonly credited with historical importance. Seeking out those experiences in various locations of inter- national practice increasingly means that researchers take to the field to find the women and study their experiences with war and other artifacts of international relations in situ. That study-up practice has begun to make ethnographic methodol- ogy acceptable, indeed it is now one of IR's normal research approaches (Ackerly *et al.*, 2006). Experience is a complex phenomenon, however, free neither of abstract elements nor of contested meanings, wherever it is studied.

Judith Grant (1993: 30) describes how experience emerged in early feminist theorizing as *the* measure of oppressive situations for women: "oppression included anything that women experienced as oppression." There was no point in appealing to objective truth when reason and reasonableness were always defined in ways that elevated men and denigrated women. Experience-based knowledge, which became known as "standpoint epistemology," made the key slogan of the Western women's movement possible: the personal is political. Yet Grant was able to see that "the conceptual appeal to experience made it theoretically difficult if not impossible to discount the opinions and/or actions of any given woman" (pp. 31–32). It also made it difficult to decide on authentic or reliable indicators of experience among identity repertoires that would mix gender identity with race, class, generation, culture, and religion. Feminism's contemporary emphasis on intersectional analysis, which builds mixed influences into compound or hyphenated identities and experiences, is one device for dealing with complicated experiential backgrounds. Yet Hirsi Ali disappointed many Western feminists who expected her experiences of race and religion to bind her to the very communities she was criticizing. Instead, Ali identified and bonded with the social and political experiences associated with her new European home country. The weight she gave to a cosmopolitan identity led to charges that her views were dangerous, colonial, and inauthentic. Bluntly put, Western feminism wanted her to live her life a certain way. They wanted her to give more weight to her Muslim background than to her European experiences and were disappointed when *their* priorities were not *her* priorities.

Reactions to Ali show the limitations of feminist intersectional analyses and the problems that poststructural emphases on in-betweens can run into when a woman actually chooses to dichotomize and to say, in effect: "I am this now, not that." In-between locations and experiences of identity sound promising; in practice, some people's identities in this globalized time can become so multiple that they choose to sort through them and make decisions that upend some feminist and critical poststructural emphases. Then, the political realm can become very personal, trumping experiential complexities, as Ali's story attests.

Sandra Harding (1986) and Kathleen Ferguson (1993) did anticipate this problem – the tensions between woman's standpoint, women's standpoint, and feminist standpoint. At issue was the question of whether any woman's experiences could be accepted as an indication of where oppression lies. Is it not likely that women's experiences will be damaged by a history of subordination? Can their recounted stories therefore stand or must they be mediated by feminist knowledge? And whose version of an experience is the right one? These are questions that social anthropologists confront all the time: Carolyn Nordstrom (2004), a well-known anthropologist who investigates the experiences of war, ran into considerable false information while she was studying communal violence in Sri Lanka. Harding (1991: 123) came down on the side of feminist mediated knowledge, saying "it cannot be that women's experiences in themselves or the things women say provide reliable grounds for knowledge claims about nature and social relations. After all, experience itself is shaped by social relations." Poststructuralist Ferguson preferred to question all fixed meaning claims, including any single feminist stream of mediation, asking instead how power is manifested in the gender stories that conventional society rehearses and in the substitute stories feminist standpointers develop. Lauren Berlant (1993: 571fn) further complicated the issues by suggesting that experience is something that is "produced in the moment when an activity becomes framed as an event, such that the subject enters the empire of quotation marks, anecdote, self-reflection, memory." More than a category of authenticity, "experience" in this context refers to something someone "has," in "aggregate moments of self-estrangement." It is a position that could show Hirsi Ali's vehement objections to Islam, and the experiences it produces for women in general, with related to events of her own self-alienating history.

My suggestion during the early debates on feminism and experience (Sylvester, 1994b) was to find research fulcrums of standpoint and poststructuralist thinking by assuming that standpoints will be multiple *and yet* people will carve out leading identities for themselves experientially, in ways that may or may not fit received categories. This approach enables researchers to investigate identity components while also recording and interpreting the experiences they hear, since, to quote Ferguson (1993: 27), "interpretation of various kinds is all there is." Indeed, the experience problematic has largely shaken out this way among the latest generation of feminist IR researchers. Experience is no longer seen as a compass pointing toward any true meaning of events or feelings and more as a methodology that enables research to focus on ordinary people and ways that international relations affects them and is affected by their actions. It is an empirical approach open to the many surprises a researcher finds in the field about herself, her own standpoints, and those of people she interviews and observes. The point is to gather experiential information from parties to and participants in war, not to judge whether they are doing the right or wrong things, whether they are feminists in actuality or spirit, or whether they can accurately identify the problems and opportunities around them. The researcher can thereby avoid the with-us-or-against-us thinking that has hampered understanding of women whose differences do not feature on the menu of acceptable feminist activities and positions of the moment.

Still, this approach overemphasizes the idea that experience is an identity tag in the sense that experience produces identity and is produced by it. Identity is a second-order phenomenon that relies on the existence not only of social mediation but also, and fundamentally, of bodies that act, process, name, own, and self-alienate. Where is the body in all this discussion of experience? Do women have access to their own experiences at all, immediately, or only after some time passes? And do their understandings of their own experiences change over time and when contexts of history change? That topic is explored in Chapter 3 but haunts the questions we address next: women's different war experiences.

New feminist war studies

Studying war experience is now a more acceptable part of feminist IR than it was even 10 years ago, but not entirely so. Along with the lament about refocusing attention from nonviolence to violence, which a war focus entails, there is also the matter of being seen as playing IR's game, joining at last in one of the discipline-defining concerns of the field: war. It can seem that studying war is tantamount to condoning it, while studying peace validates nonviolent politics. And yet war studies is burgeoning as a camp within the feminist IR camp, on its own and in tandem with feminist IR security studies. Both ground their work in intersections of feminist theory and theories from other fields, including anthropology, sociology, psychology, and literature. One of the biggest influences on feminist IR war studies is feminist IR itself, which has emerged as a separate area in feminist theorizing and a bona fide authority source for scholars in IR. To some extent, feminist IR has had to rely on itself more when branching into war studies, as women's studies, where feminist theorizing is often generated, has shown some reluctance to add war to the list of social institutions that feminists believe they must explore.

Following 9/11, several leading feminist theorists did put together conferences, and then published collections on wars that for the most part were masterminded by the USA. Strong critiques of American military and racist imperialism appeared in books edited, for example, by Karen Alexander and Mary Hawkesworth (2008) on gender and race in the wars on terror; and Chandra Mohanty, Minnie Bruce Pratt, and Robin Riley (2008) on US imperialism and intersectionalities of gender, race, sexuality, and political economy. In the UK, Nadje Sakig Al-Ali in gender studies and Nicola Pratt (2009), who works on international relations, edited a collection on transnational feminism, conflict, and post-conflict circumstances of women in contemporary Middle East war areas. More to the point of war as experience are works edited by Caroline Moser and Fiona Clark (2001), which highlight women's agency in war situations, by Krishna Kumar (2001) on possibilities for transformed gender relations following war, and by Haleh Afshar and Deborah Eade (2003) that warn about unchanged social relations that undermine post-conflict advancements for women. Very recently, the US Institute of Peace has contributed essays on women, war, and peace-building (Kuehnast, Jonge Oudraat and Hernes, 2011). These are all important collections with some compelling individual chapters. As edited books, however – like my own *Experiencing War* (2011a) – they do not constitute a feminist research area, set of war questions or field of war studies. More focused, and in some cases less abstract, feminist probings of war, social relations, and experiences are required.

In that regard, feminist theorist Butler (2004b) has had much more of an impact on feminist IR war studies by theorizing the potential for a transnational politics based on the grieving and mourning common to wars everywhere. Her theme resonates with Tarak Barkawi's sense, presented in Chapter 1, that war is actually a sign of intensified rather than diminished social connections between friends and foes. It is a point that feminist analysis needs to take on board and unpack more. The relative paucity of war theorizing in feminist circles outside IR is noteworthy. It also shows how far women's studies as a field has moved away from theories of oppression that incorporated violent politics of liberation: Marxist and radical schools of thought had many adherents in North American and European feminist circles in the 1960s and 1970s, as well as no-holds-barred spokeswomen who had little time for liberal bourgeois feminism or breeder (heterosexual) social arrangements (e.g., Davis, 1989; Daly, 1978). Today, violence is seen as mostly inappropriate to feminist political ends.

New thinking in feminist IR complements feminist collections but also departs from past and present feminist theorizing and from IR theorizing. It does not focus on the USA as war central, for example, or on wars traceable to 9/11 only; nor does it focus on transnational feminist currents of response to war, albeit elements of such themes do scatter through the literature. Taking a wide geospatial view, post-Cold War wars are seen as having many and varied authority centers at domestic, regional, and international levels. The new work in feminist IR war studies also looks more closely at experiences of ordinary people with war and draws less on abstract theoretical explanations for war like globalization, militarization, imperialism, capitalism, or patriarchy. It contests feminist IR's previously uncontested emphasis on peace and considers situations that place women in grey zones between war and peace, conflict and post-conflict.

The new scholarship does reflect back on feminist thinking, however, by asking how best to present women's *experiences* of war. Two related types of methodologies have risen to the fore in feminist IR war research so far: interviews, and discourse analysis, broadly understood. Interview research enables the investigator to talk to the subjects of war directly, often on their own turf. Annica Kronsell (2006: 121) argues that "interviews are an important source of information because they can provide an in-depth, detailed account of how gendered practices are actually carried out within institutions as well as of how gendered identities are constructed." War anthropologist Carolyn Nordstrom (1995: 139) suggests that interviews enable a researcher to enter into the realm of lived experience, which has been a privileged location for feminist research; however, she points out that while local knowledge is crucial to understanding, "quoting local informants can mean a death sentence for them" (Nordstrom, 2004: 15), which suggests that there are many hidden dangers in the approach. It has also been said that traveling to the subjects of one's research is too often idealized as a purer way of doing feminist research than through other methodologies (Stephens, 1989). Bina D'Costa (2006: 139) offers a related concern when she notes that the concept of experience has not been critically analyzed in development studies or in the context of research on women in the South, where many of the new wars take place.

D'Costa's point was very apparent to me when I conducted interviews with over 400 women workers in Zimbabwe in the post-war years of the 1980s and early 1990s (Sylvester 2000). In that research context, it was a struggle to come up with questions about the women's self-identities and aspirations that would make sense to them as opposed to American women. Through trial and error, I learned to ask a "strange" question at the start: "Are you women and if so how do you know?" I initially wondered whether subjects who ranged from factory workers interviewed on the shop floor to women commercial farm workers gathered in the field, would "get" the question and respond to it. That apprehension disappeared quickly: "We're women because that's what men call us;" "We're tired of being women and want to drive the tractors;" "Women are treated like dogs with puppies;" "We want to attend training classes in Harare but our husbands say no." Subjects called "women" repeatedly drew attention to a gap between the experiences and identities thrust on them by their communities, and experiences they wanted to have. Asked subsequently what they would do if they could be the President of Zimbabwe, they had plenty to say; and most of the things they would do reflected directly on, and corrected, their lived experiences as women-as second-class citizens, lower-paid workers, and baby-makers. Many of the women asked me to tell them what I would do as President of Zimbabwe, a turnabout that attests to the feminist fact that "both the researcher and researched are subjects with agency" (Jacoby, 2006: 171), albeit often with very differing power statuses.

While interviewing was an accepted and honored methodology in the context of African Studies, it was and remains an underutilized methodology in an IR that has often associated interviewing with area studies specialties and with the subfield Americans call Comparative Politics. The emphasis on abstract theorizing across transnational locations of IR, coupled with the American IR preference for quantitative analyses, can reduce possibilities for one-on-one, fieldwork-based interviewing that yields modest qualitative outcomes. There are also impediments to interviewing built into today's university ethics committees, which hold that interviews on difficult topics can traumatize subjects (or endanger them) and also upset the interviewer. To conduct ethical research that takes one to a distant field that seems full of dangers requires considerable strategizing by researchers. Nonetheless, interviewing is a preferred feminist IR strategy of inquiry among the generation of thinkers that studies war, and one that contributes immensely both to their bottom-up analyses and to the methodological repertoire of IR. Interview "data" are usually employed in one of two ways: to "test" (very loosely understood) the main arguments the researcher is presenting, or to lead the researcher to locally pertinent research questions. Usually it is a combination of the two.

The second major methodological tool found in feminist IR war studies is narrative or discourse analysis of primary research sources or secondary texts. The exemplary text approach taken in this book illustrates a narrative approach using secondary texts. Elshtain's *Women and War* also works off a variety of war texts, including philosophical writings, IR theories, and film, and Enloe's *Nimo's War, Emma's War* launches from a base of media reports of four women's experiences with the 2003 Iraq war. Maria Stern (2006) presents a strong argument for the use of narratives in feminist war studies based on life history interviews she conducted with Guatemalan women on the eve of a peace agreement ending thirty years of violent conflict that would also bring Mayan women into the Guatemalan nation. She developed a narrative or "text" of meaning-constructing activities from partial life histories collected through interviews that centered on their struggles as Mayan women. The text enabled her to identify key elements across and between the individual narratives and thereby offer a range of representations of women's war experiences and concerns. Stern suggests that it is important for the researcher to bear in mind that "experience is always mediated through discourse; a narrative recalling a memory is the closest one can come to being privy to another's experience" (p. 185). She goes on to say that "a focus on the discursive, constructed character of stories, or lives, does not deny that people really live, and experience threat and harm, or safety and well-being. We act, experience, and live, but the meaning we give to our actions is continually constructed within a web of different discourses" (p. 185). Interviews on their own must be massaged into stories of experience that the researcher presents, which means that the two methodologies – interviewing and discourse analysis – can be closely related.

Three recent feminist IR war studies illustrate interviewing and narrative analysis. The first is Miranda Alison's interview-based study of female combatants in Sri Lanka and Northern Ireland, followed by Megan MacKenzie's research on women combatants in Sierra Leone, which also works from interviews. The third is Annick Wibben's study on narratives of security, which has sections on the contributions of Maria Stern and Carolyn Nordstrom.

Miranda Alison

Women and Political Violence: Female Combatants in Ethno-National Conflict (2009) is a study of women's motivations for becoming combatants in violent nationalist movements in Sri Lanka and Northern Ireland, and their experiences as warriors. The commonalities linking two locations of great cultural difference include the ethno-national nature of their respective conflicts and the significant number of women who became involved in them as combatants. Alison makes it clear at the beginning that "[t]his work, like me, is unashamedly feminist" (p. 10). That is to say, the study is geared to understanding and relating women's experiences of ethno-national conflict to feminist theories of gender and nationalism and questions of women, peace and war (p. 22). Her interview strategy is based on non-probabilistic snowball or network sampling, which comes down to introductions and word-of-mouth references in the field rather than the construction of a formal sample. Alison's open-ended, intensive interviews feature 17 LTTE affiliates and a variety of others who provided background information and expert views; and around 13 women ex-combatants plus others on both sides of the conflict in Northern Ireland.

Very attentive to feminist methodological concerns, Alison spends considerable time laying out and justifying her feminist research framework and addressing relevant debates within feminist theorizing. Although these discussions seem unnecessarily long, and can detract from the more interesting interview "data," they are a measure of the extent to which feminist analysts of war can experience the sense that they are studying the wrong thing – war instead of peace, female combatants instead of female peace activists, and nationalism and feminism instead of one or the other (p. 107). It is clear that Alison wishes to explain herself and the standards she brought to her fieldwork – openness, self-critique, context awareness, identity awareness, attentiveness to listening and silence, as well as recognizing the power relations that existed between her and her subjects – "I gained much more from the experience than they did" (p. 24).

Alison's detailed discussion of her feminist considerations takes up the better part of two full chapters. This is not uncommon in feminist-inspired accounts of field research or in books, such as hers, that have been reworked from PhD theses. But compared to James Der Derian's *Virtuous War* (2009), discussed in Chapter 1, which uses an autoethnographic methodology without mentioning that term or discussing matters of methodology at all, Alison's elaborations can lead the reader to think that feminist theory and methodology are more important to work out than the views and experiences of Sri Lankan and Northern Irish women. Alison would surely protest the idea that her feminist study in any way models itself on IR realist work, but there is a similarity on one dimension: in both, the theories dominate the people, or, turned around, the people come to life only to the degree that they illustrate or support the abstract academic arguments behind the study. Alison is also cautious about expressing emotional responses to her fieldwork in keeping with the tendency across feminist IR research to avoid writing personal emotions into research write-ups (Sylvester, 2011b; MacKenzie, 2011b). Again, Der Derian shows no such qualms: he unguardedly expresses his personal feelings at several points in his book.[6] In many ways, Alison's is the more traditional approach, even though it features feminist analysis and interviews with women combatants. That is not to say that her study is flawed within fledgling feminist IR war studies. From discussions of methodology to debates about women and peace, Alison displays a capacious knowledge of feminist thinking, albeit less knowledge of feminist IR works (Enloe, for instance, is cited but barely discussed).

The quotations from the Sri Lankan women show that the LTTE women are primarily motivated by nationalism; indeed, national identity could be more dominant in their decision-making than gender identity. Most of the women combatants Alison talked to also wanted to be seen as exerting agency by taking up violent politics, even if that agency complicated their lives, and even if they were recruited to the nationalist cause mainly to boost the numbers or to replace fallen men. Thus, says one LTTE woman, "in the warfront we never think that we are women and we are soft by nature. These disappear from our mind. In the warfront we have only our aim in our mind, our aim to get an independent nation" (p. 175). Such choices and influences fly in the face of some feminist propositions about women as the more peaceful people.

As for Northern Ireland, Alison makes the noteworthy point that unlike the LTTE, the IRA and other republican paramilitaries did not recruit women or operate under norms of gender equality. The women who joined the Provisional IRA (PIRA) in its early days report that they had to struggle for acceptance – "you know ... what's a girl doin' in the Army" (p. 187)? The women's movement in Northern Ireland could also trip up republican women activists, owing to its binary framing of the issues: nationalism versus women's rights. As one woman voiced the issue: "I would actually kind of probably be complimented if somebody called me a feminist, but I would call myself an Irish republican. And as bein' a republican I see equality as part of that ..." (pp. 201–202). Alison concludes by saying that "the most recurring theme in all my interviews ... was that of respect and, tied to this, of women proving themselves (to a *masculine standard*) to earn this respect. This I found immensely depressing" (p. 232, emphasis in original). The results could affirm women's flexibility and range of human capabilities, good and bad; but it is a mark of feminist scholarship that the author should make her point of view clear rather than implying through silence that the study is value-free.

A reader of Alison's book has an opportunity to learn two related but also different sets of lessons. One set is about feminist theory and methodological commitments as they relate to combatant women. The second set is about the experiences of LTTE and PIRA women with paramilitary liberation organizations. I opt for the latter lesson at this point in my feminist IR career; however, a beginning student of feminist IR would be well served by Alison's detailed discussion of the guiding principles of feminist research. Alison's final sentence is telling in that regard: "Although the very presence of women in nontraditional combat roles (at least in significant numbers) may be the drip of water which opens up a crack in the rock, to split it wide open requires a constant feminist assault" (p. 232).

Megan MacKenzie

MacKenzie does not start her study with feminist theory. She starts with Africa, with war in Sierra Leone, and with women in war and in the post-conflict period, 75 of whom she interviewed. (See MacKenzie, *Female Soldiers in Sierra Leone*, forthcoming 2013.) She tells the reader right away that the stories women combatants tell about the war and post-conflict efforts are completely at odds with the official UN and related agency stories about how well post-conflict

Sierra Leone is doing. The women's stories also undo the notion that women and children are the vulnerable parties in war, the targets of "new wars;" in fact, many women and children actively kill in wars like this, and if their experiences are not taken into account, a skewed picture emerges of what war is like, who participates in contemporary wars, and whose notions of post-conflict security are legitimated in the arrangements that follow and whose are ignored or assumed. Put differently, without an emphasis on gender and war, it becomes impossible both to launch successful post-conflict developments and to understand gender-related bottlenecks that can undermine women's agency (for example, international officials describe women combatants who do not to choose to participate in disarmament processes in Sierra Leone as "left behind;" they do not ask the women why they did not join in and what they would prefer instead).

Methodologically, MacKenzie's (2013: pp. 33–34, manuscript) is very much about foregrounding interviews and following their lead rather than lassoing them tightly to predetermined research interests, the requirements of research proposals notwithstanding. She says:

> I include as much of my primary research as possible . . . Where appropriate, I have incorporated larger sections of interview material in order to allow the reader to "hear" the voices of the female soldiers and other interviewees. In keeping with this approach, I chose not to edit and disaggregate all of the longer quotations into precise thematic topics. Rather than splicing and cutting and pasting each section of quotation to serve the purposes of the chapter, some longer quotations have been left in full and therefore seem less strictly cohesive, including references to multiple themes or weaving together several topics.

She makes this statement in a section of the introductory chapter called "Feminist Methodology," where, in fact, she cites more sources from feminist IR than from the larger field of feminist theory and methodology. Her selections underscore the idea that feminist IR has come into its own as an authority for field research, despite rarely appearing as such in women's studies discussions of methodology.

MacKenzie does, however, have a feminist-inspired research project. Her main theoretical argument is that women who participated as combatants in Sierra Leone's war were viewed by the international community as disrupting the conjugal order, understood as "the variable norms associated with marriage, paternity, and the family [which] involve bestowing men, through the institution of marriage, with a significant amount of power over, and access to, the labour of women" (p. 127). The threat of disrupting conjugal order meant that the reintegration of combatant women would follow a different set of standards than applied to male combatants. The women were seen as having, and were thus given, limited options. They could stay silent about their role in the war and receive some training in women-associated crafts like tie-dyeing. Or they could admit to war activities, take up training in areas that might not interest them, and risk being isolated in society and worried over by international agencies. Their contributions to the war effort were minimized either way. Moreover, international programs, such as UNICEF's Girls Left Behind, did not recognize women's agency in the war, insisting that all the women were camp followers, abductees, or victims of sexual assault; it referred to girl soldiers as "unaccompanied children" who were not eligible for the post-conflict resources that went to child soldiers (read: boys). MacKenzie found tremendous reluctance in official circles to consider girls or women as soldiers, and that created a gap in post-conflict resource provision: the boys and men received media attention and were often photographed holding weapons while the women and girls were tucked away and encouraged "to 'blend in' 'naturally' to the community and family unit" (p. 96).

That is an example of the type of deeply discriminatory thinking that feminist scholarship at large has long critiqued. Yet it recurs continuously in wars, animating post-war thinking about Zimbabwe's women ex-combatants in the 1980s (Sylvester, 2000) and turning up again in Sierra Leone a decade or so later. Despite official shifts in international and development feminist thinking toward participatory politics rather than top-down decision-making, MacKenzie finds that "women" are predetermined to be traditional, pre-war women after the war. If that assignment is breached, then their conjugally-ordered lives fall apart and, in the minds of international officers, society falls apart around them. Well-intentioned, this Western liberal mind-set constructs post-conflict processes as entirely social for women and mostly political for boys and men, the people seen as less stable and more dangerous. Little wonder so many of the women ex-combatants MacKenzie interviewed did not turn up at disarmament and reintegration centers. There was no room for "new women."

Against these gender travesties, MacKenzie's interviews show "first, that the majority of 'women and girls associated with the fighting forces' define themselves as soldiers; second, these soldiers participated in multiple and diverse roles; third, female soldiers were often perpetrators of violence, destruction, and crime as well as victims of abuses such as sexual violence, abduction, and injury; finally, these interviews indicate that the number of female soldiers was much higher than existing estimations" (pp. 67–68). As her book goes on, any sense that the usual gender binaries of war hold up evaporates. These women are self-admitted killers, not the presumed support staff or the victims in the war; their duties were the same as the men's and they showed no particular gender-based mercy when dealing with "enemy" men and women. This is a startling finding given a history of feminist IR investment in women as likely carriers of peace, which Alison discusses at length. Were these Sierra Leonean women victims of false consciousness? Did they become one with militarized masculinities? The interviews suggest otherwise: MacKenzie argues "there has been little investigation into the possibility that war was empowering for female soldiers. This is despite evidence that, in some cases, female soldiers had more access to resources, more social freedom and more political power during the official war compared with the so-called pre- and post-war context" (p. 109). The field interviews provide insights into these prevailing feminist IR questions – empirically based insights. They also present MacKenzie with details about killing that researchers in IR might not often consider, details that she found emotionally difficult to hear and to integrate into her research (MacKenzie, 2011b).

Alison and McKenzie both use interviews in their feminist studies of war; however, there is a marked contrast between the two works. One is tied closely to feminist thinking and feminist issues of methodology and research ethics. The other is inspired by feminist IR war concerns but focuses more on the country context of the study and the views of war and post-war that come out of it. Both focus on women and their social experiences of war and conflict. McKenzie finds that conservative norms of conjugal relations in Sierra Leone are brought to bear on women and girls who participated in the war – by society at large and also by some international organizations that assisted in demobilizing and reintegrating combatants post-war; such norms often operate at odds with the lessons of independence and agency that military practice has taught women. Alison takes on the larger job of comparing women combatants across two very different societies and holds them to standards of feminist thought developed mostly in the West. Both pieces of feminist war studies are exemplary in different ways.

Annick Wibben

"Staying attuned to varied everyday experience, through the telling of women's stories in this case, is central to feminists' resistance to abstraction" (Wibben, 2011: 2). The theme emerges again – women's everyday experience as a counter to the abstractions that dominate so many IR studies. For Wibben, the gathering and recording of personal narratives or stories about pertinent experiences enables investigation of the many ways the world around one gets ordered and made meaningful. Most likely Alison would agree, but she might place the emphasis on how such narratives illuminate pre-existing academic and political questions. The related but different route that Wibben's book, *Feminist Security Studies: A Narrative Approach* (2010), navigates lets narratives lead in suggesting the questions to investigate; and in that approach, her work resonates with MacKenzie's. The virtue in taking that route emerges early on in Wibben's study, when she relates a story taken from an anthropological collection (Mattingly *et al.*, 2002) of reactions by African-American mothers to the 9/11 Twin Tower attacks. The mother highlighted says that "the September 11 thing didn't really bother me" because she had too many other things to deal with in her immediate life, among them getting her kids to school and taking care of her handicapped mother who moved in with her after the terror attacks and insists that bin Laden cannot get her now because she is not alone, she is with her family. While so many in the USA rally around the flag as it sets off for vengeance in Afghanistan, these women tend to favor less violent solutions, having witnessed violence all around them as daily occurrences. Researchers would not know this type of response by starting from most IR theories.

Like Alison, Wibben grounds her work in interdisciplinary feminist theorizing allotting considerable space to elaborating the strengths and political projects of feminist analysis in the social sciences. She is less concerned to use IR as a springboard, because it ignores feminist security writings as partisan or irrelevant to the "important" security debates of the moment. She does look more favorably on IR's critical security wing, however, drawing on its interest in understanding how security meanings are constructed and how certain understandings come to dominate so much thinking in and out of IR (e.g. the idea that national security is the only proper focus of security studies). Critical IR tends to lead with post-structuralist thinkers and narrative deconstructive methodologies, inclinations that suit Wibben's purposes, despite a tendency in that camp to disregard gender issues. Wibben also draws on narrative literary approaches, offering a useful discussion of contemporary narrative methodologies that work from these elements: the centrality of the reader in constructing meaning in a text; concern to preserve rather than smooth over complexities, contradictions, and intersections in story lines; attention to historical and contemporary cultural contexts of meaning and ordering; and, from Bal (1997), the importance of identifying the focalizer in any text – the author or character who presents the views or visions of others around them and the relations established between subject and object. These elements give narrative analysis signposts and guidelines without turning it into a scientific or rigid exercise bound for the one right interpretation, story, moment, or point of view in a text. Wibben then illustrates the way narrative analysis can work in researching women in war zones, by considering fieldwork conducted by scholars who make good use of narrative analysis: Katherine Moon (1997) in Korea, Nordstrom (1997) in Mozambique, and Stern (2005) in Guatemala. As in the case of the African-American mother, the fieldwork uncorks a number of experiential views that counter official story lines. We learn that Wibben is interested in the interview-cum-narrative approaches that the two feminist IR scholars (Moon and Stern) and the war anthropologist (Nordstrom) take because they capture everyday experiences of contradictory and often shifting elements that "are not smoothed over but are explicitly acknowledged" (p. 101).

Wibben suggests that feminist methodological concerns and innovations distinguish such research from other approaches in IR, and they enable feminist IR to shift from thinking of violence and security in terms of things to be protected from harm, to narrative processes that make and unmake varying understandings of those terms. She does not return, though, to the African-American mother who is too busy to process the September 11 attacks; and she does not engage in fieldwork or present her own narrative writing. Wibben's emphasis is on the importance of narratives that might seem odd in an IR context (e.g. women sex workers around American military bases in Korea) as challenges to usual IR discussions of national security.

These three exemplary studies illustrate important facets of feminist IR research on women and war – fieldwork interviews used to illuminate propositions in the literature, and narrative analysis as a distinctive tool of feminist IR research.

Both approaches yield new information on war and the people in it. Ethnographic fieldwork, however, adds an especially distinctive methodology to IR's usual approaches. As Nordstrom (2004: 13) explains it, "ethnography must be able to follow the question. It must be able to capture not only the site, but also the smell, feel, taste, and motion of a locale, of a people that share a common space and intertwined lives. It must be able to grasp at least a fleeting glimpse of the dreams that people carry with them and that carry people to distant places of world and mind; of the creative imaginary through which people give substance to their thoughts and lives." To do all this requires more of a hands-on presence in a research site than IR usually advocates or trains its students to do. Of course, not every feminist doing IR war studies has the inclination to take to the field in this manner; nor, as Wibben's analysis shows, is it necessary in all cases to do so. I do not want to fetishize ethnography, but I do want to flag it as particularly appropriate to studying international relations up by grounding knowledge in the daily life of war and post-war people.

It is also important that the research is based in feminist theoretical and methodological scholarship to some degree. But to what degree, exactly? Alison and Wibben spend a greater proportion of their books elucidating feminist thinking than elucidating what everyday women are thinking and saying about their experiences with war and security issues. It is important, of course, to provide proper grounding for research; yet I would say that in these works, the grounding strategies reign over the subjects of the research, as though the key point is that feminist thinking must be presented carefully and with clarity to insure that it becomes accepted in IR. In a camp structured field, that is a faux problem. The real problem is that feminist groundings can be belabored, detracting from the people one is interviewing or the stories one seeks to read and analyze. In Alison's and Wibben's books, there is little sign of those people of international relations until Chapter 4 in a seven-chapter book (Alison) or Chapter 5 in Wibben's book of six chapters (the introductory quote about reactions to 9/11 becomes a one-off). If one looks back at the IR exemplars presented in the previous chapter, there is no long wind-up to their pitches, perhaps because those scholars do not have to justify their emphases the way feminists do. After nearly 30 years of feminist IR research, however, one doubts that yet another clear explanation of what feminist research is about is really necessary. It makes some sense to get to the people quickly and let the weight of that analysis carry the day rather than get bogged down in reviews of feminist literature (written mostly in the 1980s and 1990s). Move on, I say, and the work of the third exemplar, MacKenzie, does exactly that, and with admirable confidence.

A disciplinary standoff?

Six books can highlight only some of the research on war in feminist IR. Elements of the works reviewed here, along with many others, will feature in the next two chapters, which move away from reviewing disciplinary theories, methodologies, and positions on war to theorizing elements of war differently. In the main, feminist thought has been anti-war, especially when armed conflict has involved the Great Powers. Other types of wars, such as those fought for national liberation, can get more lenient treatment in feminist thought on the grounds of anti-colonial justice and historical expediency. For the most part, though, war is a zone of discomfort and ethical belligerence in feminist thinking. The tendency across time has been to say no to it, politically and analytically, rather than look it in the eye. There is also a tendency to bury war in large concepts that purport to explain its persistence, intersectionalities, and victimizations without necessarily getting the researcher near war as experience. Feminist analysis undertaken outside the IR context tends to register war questions as racist-imperialist practices of international relations, often targeting the USA and/or Israel as confirmed warrior states in the present era. Feminist IR researchers, by contrast, talk more about insidious processes of militarization that have been revved up by terrorism, local identity politics, and political economies of neoliberal globalization that feed and live off war engines.

Curiously, some habits of feminist IR war thinking overlap with IR. It cannot be said that either field has been especially or consistently attuned to war studies. IR arose in Europe following World War I and focused on understanding why wars occur, especially as international relations rocketed toward the second major conflagration of the twentieth century. Perhaps Cold War hot wars in Korea and Southeast Asia cooled IR's war interests; or perhaps it was the post-Cold War shift from war and peace to the rights of people relative to the rights of states that shifted the focus. So did events in New York City and elsewhere in 2001 and the years following; suddenly, terrorism studies, which had become the purview of a dwindling number of analysts, came stage-center with hyper-concerns about national security and its foreign and homegrown threats. IR's interests in war never disappeared entirely, as the discussion of new wars in the previous chapter indicates, but it has been dwarfed in recent years by the security emphasis across the field. John Mearsheimer and Stephen Walt have brought security issues into their studies of wars involving Israel from the perspective of domestic American foreign policy, and critical IR circles focus almost exclusively on myriad security problematics. James Der Derian delves into war without apology and without taking personal distance from war experiences. For the rest, we recall Tarak Barkawi urging IR colleagues to take up the study of war with renewed vigor, this time from historical and postcolonial points of view.

In feminist and feminist IR circles, there are parallels. A steady drizzle of pieces analyzing aspects of war appears, but very little interest is shown in claiming war as an especially relevant topic for feminist analysis (e.g., Cohn, 1987; Cockburn, 2007; Mertus, 2000). Studying war instead of focusing on peace has only come to the feminist IR agenda in the past decade and is still in its infancy. Feminist analyses outside the field of IR have largely ignored IR knowledge on war and also IR's feminist wing, coining its own studies of transnational connections around issues of race, class, sexual preference, violence, and gender. When taking up war issues it is odd to find that feminist scholars outside IR can focus on causes and practices of war, like IR has done historically, and not on war experiences. One exception is Butler, whose unusual work within the feminist canon overlaps with Fierke's IR constructivist concern with war trauma politics and is treated in Chapter 4.

Major differences between IR and feminist IR come to light over disparities in research focus and methodological approaches. IR does not conceptualize international relations as encompassing ordinary people and their experiences with the actors and processes it takes as canonical – states, markets, militaries, international organizations, security, development and so on. Feminist IR does, and some (not all) feminist analysis outside IR also does. Feminist IR studies of war are especially distinguished by a bold interest in presenting the field with ordinary women in war, often taking to locations of conflict to speak to those women directly. This chapter has foregrounded two studies that do so and one that considers results of interviews and textual analysis – narratives – as key sources for feminist security studies. IR operates at different levels of analysis than feminist IR, and that difference can account for most of what feminists doing IR lament as ships passing in the night (Tickner, 1997) or IR willfully ignoring feminist IR scholarship. IR clearly does not sideline empirical research on gender and conflict if it uses aggregate data related to the advancement, or not, of women internationally (Hudson *et al.*, 2009; Caprioli, 2003), or features gender-based research on groups most likely to be injured during wars (Carpenter, 2003). Those studies use scientific methodologies that IR does not find troubling.

There is an extensive feminist literature, however, on gender biases embedded in scientific approaches to research (e.g., Harding, 1986), not least the mistake of conflating gender and sex and analyzing the data without considering "the symbolic meaning that creates social hierarchies based on perceived associations with masculine and feminine characteristics" (Sjoberg, 2010: 3). There is also considerable early feminist IR literature on gender biases underlying IR theories and research that derive from philosophical and historical sources (e.g., Sylvester, 1994a; Tickner, 1992; Peterson, 1992). Alison and Wibben review many of these considerations in the lengthy sections presented to justify the different methodological decisions they make as feminists doing IR research. MacKenzie's feminist IR analysis takes much of this discussion as given and does not repeat it before confidently declaring that her work is based on interviews and analysis of them. Her methodological focus resonates with Shapiro's gender-based analysis discussed in the previous chapter, where war comes into some women's home spaces and the home spaces enter war's political spaces. It is important to point out, however, that how men and women are categorized or understood in feminist research might have made little difference in Sierra Leone and other post-conflict sites, where people are assigned locations based on what agencies make of them and their assumed home spaces.

The next two chapters move away from surveying war approaches in IR and feminist IR to broadening some of the promising ideas about war and experience that have been noted so far. The argument I want to make is that war not only provides experiences to people, war is experience and can be studied up from bodies. Starting with the body departs the levels of analysis commonly employed by IR and enables a wide range of space–body relationships to be investigated, including those involving people who do not become actual combatants or victims in wars. The next chapter considers the physical experiences of war, while Chapter 4 takes up related issues of emotion and war experience.

PART II

Rethinking elements and approaches to war

3

WAR AS PHYSICAL EXPERIENCE

What if we were to turn the study of war around and start, not with states, fundamentalist organizations, strategies, security issues, weapons, and not with the aim of establishing the causes of war, as has so often been the case in IR? What if we were to think of war more along the lines of feminist-inspired studies presented in the previous chapter, where war is a set of experiences that everyday people and elites have physically, emotionally, and social-ethically, depending on their locations inside and even far beyond war zones? To move in that direction would be to admit that war is a social activity of collective violence around which a wide variety of bodily experiences are created, altered, and can themselves create or constitute war. The works shown in Chapters 1 and 2 indicate that there is sufficient interest in war and experience across IR – direct interest and submerged but unavoidable interest – to warrant the suggested addition to more abstract war studies. Clearly, IR's feminist wing of war studies, which is still taking shape, has implicitly made that kind of proposition a touchstone of its research, as have critical IR scholars like Tarak Barkawi, Shane Brighton, and Michael Shapiro, and Karin Fierke from the constructivist wing.

To study war as experience requires that the human body come into focus as a unit that has agency in war and is also the target of war's violence. The body could be an individual or an aggregation of bodies with differential spatial relationships to the actions of war. Bodily experience can seem a distinctly unIR-like place to begin, despite the many hints across the IR canon that war is peopled and not just an abstract politics and technology of extreme disagreement. That the IR canon in general has not been directly interested in the human and its body is evident in its extractions of a few bodily attributes it finds useful and the habitual exclusion of the rest of the corporeal from most of its theories. One prime example is human rationality, which features time and again across the discipline, as we have already noted. Mostly it is drained of body–mind and body–emotion elements and made parsimoniously logical (Bueno de Mesquita, 2010). Some dropped bits are picked up by scholars of political psychology (e.g., Kinnvall and Nesbitt-Larking, 2009), but the gendered body can be overlooked in historical studies of foreign policy "decision-makers" (Snyder *et al.*, Kinnvall 2002), despite the strong presumption suggested in the works that decision-makers are bodily male (Sylvester, 1994a). There has also been some forcing of gendered bodies into male/female frameworks (e.g., Caprioli, 2003; Carpenter, 2003), which makes for easier data manipulation but cannot capture nuances of social practices and conditioning. Some critical IR has even been known to relegate certain bodies to footnotes rather than put them into the main body of their texts (C.A.S.E., 2006; Sylvester, 2007), or to place emphasis on men and, implicitly, their bodies operating in military circumstances (Der Derian, 2009). By contrast, the feminist IR analytic tradition is attentive to the body and its centrality to social institutions, increasingly including the institution of war.

This chapter sets the stage for thinking about whom/what is experiencing which kinds of bodily actions and reactions related to war. It focuses on the human body as a physical fact of war, a performative arena, an imagined presence in war, and a key target and site of collective violence, working with ideas that appear across IR camps and in related disciplines. The chapter leads with some of those related literatures, starting with Elaine Scarry's contention that war is about injury to bodies, and moving to debates by feminist and some IR scholars on bodies as physical, symbolic, and performative entities, featuring Judith Butler's ideas. It then returns to literatures specifically addressing what happens to bodies in war as analyzed in two exemplary feminist studies, one focusing on the Balkan wars (Zarkov, 2007) and the other on bodily experiences of war in the Congo (Eriksson Baaz and Stern, 2010).

War is about injuring bodies

People live in and among bodies and those bodies influence people and social institutions such as war. How the forward and backward body–society linkages work with respect to war, however, remains murky. There is an unmistakable reluctance in fields that study war to talk about the obvious fact that the institution of war is about preparing some bodies to injure some bodies and to safeguard others. Elaine Scarry (1985: 64) insightfully notes that "one can read many pages of a historic or strategic account of a particular military campaign, or listen to many successive installments in a newscast narrative of events in a contemporary war, without encountering the acknowledgment that the purpose of the event described is to alter (to burn, to blast, to shell, to cut) human tissue, as well as to alter the surface, shape, and deep entirety of the objects that human beings recognize as extensions of themselves." Feminist analysts have also remarked on the ways that military organizations use clinical languages to distance themselves from body-injuring activities of war, bodies that injure during war, and bodies that experience war injury (Cohn, 1987). Injury is the truth of war hiding behind the common war-defining term of "collective violence." Injury may not be the only way that bodies experience war physically, but it is a central objective of war and its technologies. Indeed, as Scarry (1985: 67) puts it, "reciprocal injuring is the obsessive *content* of war and not an unfortunate or preventable *consequence* of war" (emphasis added), which is the impression one can get from IR.

Despite an increasing emphasis in Western ways of warring on computer-guided air wars, aimed at achieving calibrated objectives with the minimum of "collateral damage" to people and infrastructures, the targets are bodies of individual enemy operatives. At the other end of the technology scale, so-called new wars of our time often feature rape, machete attacks, and armed children, which indicate clearly – more clearly than Western war planners would do – that bodies are both the perpetrators and the targets of today's wars. Americans knew that during the Vietnam War, when the US government regularly reported "body counts" to the media. The UN knows this today by directing increasing resources to understanding and preventing the epidemic of rape in certain war and conflict areas of member states. In either case of high-technology wars or low-technology wars, material bodies remain at the center of war, whether that is acknowledged and built into theories of contemporary war or not.[1]

That Scarry could write about bodies and war in this way suggests that she has not been an IR-imprimatured specialist on international relations. She is currently Professor of Aesthetics and General Theory of Value in Harvard University's Department of English. That location in academia reveals much about the war smokescreens of IR: the field that claims war as one of its core areas of research has largely placed itself outside the range of war's fire by focusing mainly on abstract causes and strategies of war and security. Distance can be maintained even when a scholar has trained for war, as John Mearsheimer (2002) did at West Point, and even though he notes in personal comments that "West Point taught you to suffer." Scarry (1985: 64) offers three reasons that the body goes missing in much scholarship on war: "omission may occur out of the sense that this activity is too self-evident to require articulation; it may instead originate in a failure of perception on the part of the describer; again it may arise out of an active desire to misrepresent the central content of war's activity (and this conscious attempt to misrepresent can in its turn be broken down into an array of motives, some malevolent, some relatively benign)."

Reason three – misrepresenting the central content of war – jumps out as I read Scarry's work. George W. Bush was reluctant to allow the media to photograph the "body bags" and coffins of American soldiers killed in Afghanistan and Iraq. Despite injury being the central content of war's activities, and despite the fact that Americans were dying in the wars on terror, Bush endeavored for a while to effect a propaganda disconnect from those facts. The point was to keep the American electorate on his side, the side of interventionist war, by controlling perceptions of American war capabilities. The USA might be the only country that wants to believe that war is solely about deaths of unknown people "over there". This myth is fed by the use of advanced air technology, which can unleash injury on other people and architectures while insulating many nonground-based American soldiers from reciprocal injury. The failure to say much about bodies of dead Afghanis and Iraqis also illustrates the first reason Scarry gives for bodies that go missing in war analysis: apparently, those deaths are self-evident and do not require articulation. Under the Obama administration, this principle has been reversed: the killing of well-known enemy individuals has become something to boast about, underplaying (a form of misrepresentation) the many times that faulty intelligence has resulted in drone aircraft targeting innocent families celebrating weddings or just sleeping at night in their homes.

Scarry relatedly notes that "even in a relatively confined war the events are happening on a scale far beyond visual or sensory experience and thus routinely necessitate the invocation of models, maps, and analogues" (p. 101). She is referring to the distancing tools that turn planned injury into rational games or mathematical formulas; although it could be extended to situations where people experience war indirectly and at a distance through TV maps and analogues. Still, if we stay with her logic, it is important to recognize that war is not an injury politics that can always be directly seen, felt, tasted, smelled, or heard even by bodies in war zones. Ordinary people draw on maps and models in their heads, too, as they try to figure out where the latest danger lies and the routes they should take to escape approaching killers or air bombardments. That is to say that part of the physical experience of war entails gauging how remote or close "it" is to oneself on any given day and moving one's body accordingly. Tracy Kidder (2010) tells the tale of Deogratias, a young medical student in Burundi, who miraculously goes undetected by genocidal gangs who invade his hospital one morning in 1993, seeking to kill Tutsis like him. He spends months running and hiding in the bush around Burundi and Rwanda, always barely evading the forces of genocidal war and their stark objective of physically injuring – killing and maiming – people with the "wrong" bodies. His internal maps and models of survival keep him alive.

Yet as with many others escaping war zones of the world, he finds that his stock of useful maps and analogues runs out when he becomes a refugee in a distant place. In Deo's case, the place is New York City, where he endeavors daily to map an alien location so he can find his way to a new life. In one scene, Deo tries to decipher where he is and where he is going on the New York subway system. The official subway maps do not help him much because they operate in a language and on a scale that is outside his experience:

> Deo rode the subways from one end of the line to the other, again and again. He studied the maps on the walls of the car. They were hard to read ... he realized that a map was no good to him anyway, because he had no idea where he might be situated in the multicolored lines and foreign words and symbols ... A couple of times he disembarked and found himself surrounded by cars and people rushing by in all directions and by buildings so high he had to search for the sky, and, feeling even more lost up there than on the trains, he went back underground ... (P. 11)

Deo sleeps for a while in New York's Central Park, a place he finds far more familiar and safe than the various flophouses where he is directed to stay. His body, seriously weakened by months of strenuous effort to hide from killers in Burundi and Rwanda, further deteriorates as he struggles to survive in an American zone of peace with maps that make no sense to a body carrying the experiences of distant war.

One might expect the critical wing of IR and security studies to zero in on situations like those that Deogratias and so many others in and away from war zones face, particularly since biopolitical analysis is a favored approach in critical circles. Following Foucault, whose writings are the bases of this type of analysis, Mick Dillon and Julian Reid (2001: 41) characterize biopolitics as "the kind of power/knowledge that seeks to foster and promote life rather than the juridical sovereign kind of power that threatens death." Less positively, one could also say that biopolitics has to do with the administration of lives and livelihoods via discursive "rules" that establish and regulate bodily activities such as birth, death, gender, marriage, work, health, illness, sanity, rationality and so on (Sylvester, 2011d, 2006). Biopolitical analysis generally focuses on the power of life and death held by bona fide states, and its reference points are usually Western states that operate most of the time according to established laws. Biopolitics could also focus on ordinary bodily lives and deaths in war zones, an approach that would be especially germane in cases of so-called failed states, where official authority is weak or nonexistent, as is often the case in contemporary war areas outside the West. In such situations, discursive rules would still operate about who may kill, be killed, or enabled to live, but these would not issue from, or garner their authority from, state elites. They would be less obvious and accessible to outside analysis. The rules could be set by gangs, guerrilla fighters, mercenaries, private security forces and the like. The shaping of those biopolitical rules, and the consequences of them, however, would involve local input and would also target ordinary bodies.

The possibility of considering the biopolitics of ordinary people's lives and deaths in wars is especially latent in the theories of Giorgio Agamben (1998). Agamben considers circumstances in which states make exceptions to their own laws about killing to create possibilities for some to be placed outside any juridical or discursive rule. The death of soldiers in state-run wars comprises a common exception to usual laws forbidding killing in the domestic arena. Soldier deaths are honored by states and treated as sacrifices of life for a larger good. When the state holds people incommunicado, gathers people into asylum facilities or extermination camps, or makes people disappear without public acknowledgement or eulogy, exceptions are being taken that end up excluding some people entirely from the community of rights and from any hope of justice. Those who exist under such conditions live a bare life-existence in a "zone of indistinction between sacrifice and homicide" (p. 83). They are, in Agamben's terms, "homo sacer."

Agamben's work is insightful for understanding contemporary behind-the-scenes activities of warfare, such as detaining people indefinitely without trial on some suspicion that they are terrorists, torturing and demeaning such people, as took place at Abu Ghraib, and maintaining the Guantanamo holding camp for years on end. Bodies in such circumstances have no recourse to state protections, means of assistance, or hope of release. Yet Agamben's analysis does mirror the tendencies of IR to focus on the activities of the state and not the bodies of those injured by the state. His emphasis on the exceptions that sovereign states take could be turned around, however, to put the spotlight on the bodies and lives of people in bare life-contexts of war. That is to say, one could foreground the people experiencing bare life-situations in wars relative to situations of sacrifice, rather than theorize mainly about the (Western) state and the life and death it can exact even under the banner of democracy. To do so would be more in line with the idea of war as bodily experience, instead of something that should be abstracted onto a "higher" plane or level of analysis, as one often finds in critical IR. It is also important to recognize that power is not fully monopolized by sovereign authorities. As Foucault was at pains to have us understand, power begets power, which is to say that power and authority have many locations, in general and surely in war situations.

That point often comes across clearly in analyses of war and conflict situations that look for nonstate forms of power, authority, and bodily agency at fulcrums of biopolitics, development studies, feminist studies, and postcolonial analysis. Some nexus points of development studies and postcolonial studies, two cognates of IR that are often relegated to the field's margins, draw on Agamben's concept of bare life (Sylvester, 2006, 2011d; Kapoor, 2008; Raghuram, Madge, and Noxolo, 2009; Biccum, 2005). Megan MacKenzie's (2011a, 2013) work on female soldiers in Sierra Leone's recent wars shows what a more complex fulcrum of feminist, postcolonial, and biopolitical analyses might look like and reveal. She starts with the point that war is usually presented as an exceptional time and politics when normal laws and norms are suspended. Yet MacKenzie finds it odd that in such situations of exception, societies, international organizations, and scholars continue to operate under the "assumption that men and women have standardized experiences of war" (2011b pp. 65–66). Men are the standard soldiers. Women and children are the standard victims rather than perpetrators of injury, overall and in the case of civil war in Sierra Leone. Agamben seems to share these commonplace views, which would be called into question if he and others were to study up from people and their bodies as well as studying down from the state onto an analytically homogenized "population" of abstract people.

That failure is the result of disregarding particularities of bodies, including their gender, and refusing what the postcolonial analyst and anthropologist Veena Das (2007) calls a "descent into the ordinary," the everyday, the mundane, and private memories to understand events of violence, war, and injury. Das thinks there is suspicion of the ordinary in the social sciences, which shows up in theoretical impulses to frame agency as something that enables escape from, rather than descent into, everyday relationships for meaning. By contrast, the ordinary is what postcolonial studies of concrete and literary situations emphasizes. Taking a postcolonial anthropological perspective, Das studies the brutalities and women's memories of such surrounding Partition in South Asia. She discovers that certain basic facts of that violent period have been airbrushed by the Indian state in its discourses of nation-building. For example, the abduction and rape of women gets considerable attention in this discourse while violence to men is ignored, leaving the impression that the nation is a masculine stronghold and not an entity of men *manqué*. It also results in post-Partition laws mandating that women who were abducted into communities other than their own must be returned to their families, which effectively means returned to the authority of the men in the family.

Descent into the ordinary is something that postcolonial novels or novels depicting postcolonial wars can do rather well. The one story of war that has had the most impact on my own thinking is Gil Courtemanche's (2003) *A Sunday at the Pool in Kigali*. Courtemanche is a Canadian writer whose Rwandan characters "existed in reality" (preface), as he knew them on the ground in a country building up to a biopolitical catastrophe. The novel features three identity and corporeal groups: Western embassy and military personnel, clerics, and development/aid workers; Rwandan hotel workers, militants, AIDs sufferers, taxi drivers, prostitutes, lovers, wives, and families; and the Rwandan state and parastatal militias. The Westerners are opportunistic and perniciously biopolitical. They are there mostly in a helping capacity, and they also help themselves to a full range of local bodies. The diplomatic community in Kigali consists of hard-drinking white bodies that gather around the swimming pool at the local hotel or on the golf course. Imbibing B-movie desires and ambitions with their French food and wine, they mythologize the often brutal sex they have or want to have with locals and generally make a lot of noise. The few European soldiers about are either oafs or inefficacious standabouts guilty in Courtemanche's eyes of every colonial trope in the imperial book of manners.

The Rwandans are no angels, but many play an intricate ethnic game of "concealment and ambiguity with awesome skill" (p. 10). In an environment where a particular body-linked ethnicity is good one day and bad the next, one response is to turn oneself into whatever seems to be privileged at the moment. Alternatively, one can try to lose ethnic identification and find some zone of distinction, some place of safe citizenship that will safeguard the body against the biopolitics of the present moment and the foreseeable turnabout politics of the future. As the personal collapses into the political through the body, though, no camouflage really works. Everyone sits anxiously in the forecourts of genocide trying to ignore the scent of extermination. Yet Courtemanche shows people also living daily life as though nothing were amiss. No matter that trucks of youths brandishing machetes cut down people at random and leave their bodies by the side of the road to be found anew every morning. Life continues. Meanwhile, the Rwandan state has a virulent presence but little corporeality. "It" consists mostly of disembodied voices – radio broadcasts that call for the demise of the Tutsi "cockroaches." The key players of state are nonstate, which is to say they are paramilitary extremists who plan genocide against the minority Tutsi population and tell everyone about it well in advance, as though advertising an upcoming social event. Sovereignty lodges not in any political institutions or laws in this Rwanda but in Foucauldian governmentalities of bodily control that make the exception to the usual rules against killing the rule.

One character in the novel stands out amid an excellent cast: Methode. He is experiencing bare life most acutely, presciently, and complexly as a 32-year-old AIDS victim. As other forms of corporeal violence swirl around him, Methode knows he is going to die either way – from the chop of a neighbor's machete or from international biopolitics that keeps the best AIDS drugs from Africa in the mid-1990s. Deciding it is better to die of a disease he got from good living than in "the cataclysm of the poor ... with bellies ripped" (p. 41–42), he takes control of his death. Methode hosts his own farewell party in his hospital room, complete with sex, morphine, whiskey, and the companionship of good friends. He then dies "fulfilled, if only through his eyes" (p. 45). His mode of departing in celebratory style – emphasizing the body as a vessel of pleasure and choice – substitutes another biopolitics of exception for the one that was to be his fate. It is a bad trade-off, but one that lets Methode redefine boundaries within which he is not as thin as his arms suggest or as dead as media depict him. Methode is way beyond Agamben, projecting agency into bare life.

In another time and place, MacKenzie finds that three-quarters of the Sierra Leonean women soldiers she interviewed reported having combat-related duties during the war, ranging from amputating the limbs of enemy bodies to planning warfare strategies. These duties were too gender-exceptional – one might say beyond the realm of exceptionality – to be given adequate and respectful recognition in local Sierra Leonean tradition. But in addition, women's war duties were quickly put on the shelf by international agencies that saw their mission as reintegrating combatant women into "domestic, peaceful, private life ... that would happen 'naturally' with time ..." (2011: 75). MacKenzie argues that Agamben neglects gender considerations so consistently that it compromises his biopolitical theory and feeds a larger, equally neglectful tradition in international theorizing. She says: "Agamben is determined to discover how exceptional violence is possible in the face of legal mechanisms and established order; yet if he were to take gender seriously the answer would be apparent. The construction of the domestic, private, and peaceful sphere (with helpless and innocent women and girls at the centre) gives meaning to any idea of exceptional politics and provides the justification, the *raison d'etre*, for exceptional violence" (p. 76). Thus there is Methode, who is making his own last decisions admirably. There are also, however, scores of women who are possible victims of his sexual appetites. Their names we do not learn. Their deaths go unremarked in the novel.

Das's parallel line of thinking is that certain practices that enter the lexicon of social normality – such as returning abducted women to their homes (men) – can be pathological though entirely lawful. The practice could force women into situations that lead to bare life, when what women require or may seek are situations that enable them to live life. If the focus is, instead, on "making the everyday inhabitable" (Das, 2007: 216), a displaced woman "could occupy the space of devastation by making it one's own not through a gesture of escape, but by occupying it as the present in a gesture of mourning" (p. 214). The idea that both MacKenzie and Das explore, and which also appears in the Deogratias and Kigali stories, is that wars injure, and escapes from war zones or their memories can also injure. But post-conflict settlements that rest on space–body relationships that return women to the *status quo ante* war, in the name of justice and the law, must also be evaluated as possibly contributing to war injury. Otherwise, the biopolitical tales of war, post-war, and decisional trade-offs might carry on a gender tradition of giving all the meaty roles and perks to the bodies of men.

Yet the physical body is a contested entity

Of course, the preceding analysis presumes that we all know bodies – what they are, what they look like, what their gender is or culture or age, as the thing injured in or seeking agency through war. In fact, there is no consensus on what a body definitively is, even in feminist circles, where the body has been discussed and theorized at length. It has been commonplace in feminism historically to talk about the body as primarily a sexed and gendered physical and social entity, distinguishing between sex as fixed body biology relative to gender as the social meanings and practices associated with biological-reproductive differences. That physically sexed body is a material and visible thing that comes in male and female forms, conventionally. Gender, by contrast, refers to stories repeated, social assignments habitually accorded, and identities embraced, or rejected by, physically sexed people. Today, queer literatures have made it clear that sex and gender are far more complicated "trans" processes that extend beyond the usual men/women dichotomy. In the social sciences, however, it is still common for researchers to mistakenly read sex off the body or the proper name, and then compound the miscategorization by referring to those bodies using gender terminology – men and women. In fact, people who look like men may not have male sex organs, and being sexed male may or may not have anything to do with that person's enactment of gender. Ditto for "women." Judith Butler (1993: 2), one of feminism's leading theorists of sex and gender, says bodies "never quite comply with the norms by which their materialization is impelled." There is always a zone of the physical that lies outside the usual body-based rules of sex and gender and can render its inhabitants social outcasts, deviants, or traitors to their "sex."

For most feminist analysts today, the seemingly reliable category of body sex is definitely problematic; it is seen more that the sexed body is a norm, expectation, and performance rather than a set of fixed characteristics. Moreover, the sexed body might be produced by the power of regulatory regimes of truth that govern social life anywhere. Normed performances of sex would then exert strong conformist power for bodies and also produce bodies that do not conform, do not recognize their place, do not act socially as they "should." Such bodies can be shunned, "cast off, away or out ... within the terms of sociality" to invisibility or political exile in unlivable, unvalued statuses lacking proper subjectivity (p. 243n). Thus under-stood, bodies are neither entirely fictional things – they do have materiality and experiences – nor are they fixed entities with determinant characteristics, sub-jectivities, and modes of behavior. But there is a considerable range of thinking on this.

Rosie Braidotti (1989: 93), whose work represents more of a European than a US take on the body, has argued that it is important to "re-connect the feminine to the bodily sexed reality of the female, refusing the separation of the empirical from the symbolic, or of the material from the discursive, or of sex from gender." That reconnection centers on complex issues of sexual difference that she does not see as physical, material, or performative per se. Her view of the body, then, suggests a sexed entity whose existence is shaped by symbolic rather than strictly physical attributes – or rather, the material and the symbolic intertwine. The phallus is a key symbol in Braidotti's framework. There are phallic expectations of abstract virility for men and less powerful and socially esteemed – and more volatile – expectations of women, who do not possess phallic power. Other points of view abound. Elizabeth Grosz (1994) argues that perceptions of social achieve-ment historically are effects of sexually specific bodies that become implicated in power structures. Sexual difference is a mobile concept that is "able to insinuate itself into regions where it should have no place, to make itself, if not invisible, then at least unrecognizable in its influences and effects" (Grosz, 1994: ix). That is its power. Susan Bordo (2003: 288), by contrast, maintains that "the very notion of the biological body is itself a fiction" at a time when bodies can be so commercialized that they "become alienated products, texts of our own creative making . . ."

War is one institution and set of locations where conventional norms of sexual difference do seem to reign almost unfettered, at least in wars where uniformed state militaries have confronted each other on the ground, air, and water. War is an obvious phallic location of bodies sexed as male, irrespective of how such people might view their own gender. One can also say that the norms of war require that official perpetrators of collective violence engage in certain male-identified experiences that Cynthia Enloe repeatedly calls militarized masculinity. Bodies of sexed women are meant historically to be outside the realm of military engagement, says Jean Bethke Elshtain, and left either cowering from bombardments and invasions or safely on the home front working and waiting for the men to return alive or dead. Of course, the all-male militaries of most states are a thing of the past, as are wars that rely on formal militaries per se. Yet the struggles that have taken place around the admission of women to armed forces, and especially to combat operations, show how fixed the idea can be that certain bodies may do war and other bodies may not; ditto for bodies that are sexed male but are sexually interested in other such bodies rather than in female bodies: only recently have they been named and officially accepted as bona fide men of war in the USA.

That the transhistorical and transnational institution of war works mostly with sexed bodies of men emerges as an empirical fact that the feminist IR scholar Joshua Goldstein's monumental study of *War and Gender* (2001) mostly confirms. With only a few historical counter-examples on record, he finds that the institution of war is sex- and gender-marked as the territory of heterosexual men who can *allow* sexed bodies of women and of gay men to join official state militaries. MacKenzie's research suggests that the same norm exists outside some contexts of state-run wars: men fight and women can be allowed and even encouraged to do so, too, but only as an exception that must be remedied as soon as the war ends. Women combatants then come under pressure to erase their war experiences as though these never existed, rather than wear their war labor with the honor accorded sacrificial killing and dying, to use Agamben's terms again. Sexual difference is expected as a rule, a body, a norm, and a recognizable empirical reality. Yet Goldstein also makes the important point that "culture directly influences the expression of genes and hence the biology of our bodies . . . no universal biological essence of 'sex' exists, but rather a complex system of potentials that are activated by various internal and external influences" (p. 2).

Carrying this point forward exposes cultures of heterosexuality and race that have shaped the ideal warrior in the Western context as male, white, and not-queer sexually. It is a subset of body norms that Shannon Winnubst (2003: 6) summarizes as the larger cultural standard against which all physical bodies are determined, measured and actually made to inhabit inferior spaces of otherness:

> . . . the white male heterosexual body erases its own corporeality—its own particularity and specificity—so that it can enter into the totalizing realm of the universal. In turn, it assures its strict and rigid boundary from the Other by erasing all otherness—to be Other is to be in the body, to be particular, to be less than the universal, to be flawed, limited, marked, different. And this difference makes no sense without any access to that guarantor of meaning, the universal, which in turn necessarily erases any difference—any particularity, any body, any Otherness. A perfectly sealed circle of self-reflecting, solipsistic Sameness—or what has passed as "truth." The boundaries are thus maintained through whiteness' and maleness' and heterosexuality's invisibility.

Bodies that cannot match those ideal-typical characteristics are doomed to lumber along in (visible) bodies, weighed down in the variety of ways that feminist analysts have articulated across the decades (e.g., de Beauvoir, 1957, hooks, 1990 and Bordo, 2003). It is an argument that acknowledges the nonuniversality of bodies, but that announces a universal standard for one sexed, racialized, and sex-engaging body-type wherever it is found. And accompanying that standard is a range of identity and characteristic practices, which are questioned, mocked, parodied, accepted and resisted – in fact and in spaces of the imagination.

In feminist circles and science studies, Donna Haraway was one of the first to suggest that bodies can be cyborg, which she describes as a combination of "text, machine, body and metaphor" (1991: 212). That is to say, bodies can be part human, part culture, part technology, and part stand-in for other resemblances and meanings. She later expanded her thinking to propose an understanding of bodies in today's world as ensembles of living and technoscientific connections, saying that "a textbook, molecule, equation, mouse, pipette, bomb, fungus, technician, agitator or scientist, can—and often should—be teased open to show the sticky economic, technical, political, organic, historical, mythic, and textual threads that make up its tissues" (Haraway, 1997: 68). With boundaries shed, blood and screens conjoin, and cursors and cursing together form not only artificial intelligence and life as robots; all too often they produce multi-vectored life as cyborg anxiety. Boundaries breached mean that contaminations have many forms – computer viruses and phishing, social networking, air-borne disease, mixed(-up) blood, tattoos, gay marriage, other-penetrated nations, and the stranger danger of killers queuing to board your razor-sharp, hyper-engineered, securitized flight. And that is all before one gets into the serious possibility raised by Jane Bennett (2010) that the nonhuman things can be actants.

Which brings us to vampires, zombies, werewolves and other undead nonhuman-human bodies that roam the landscapes of Western imagination, much as spirits of various kinds roam other cultures. There is tremendous fascination of late with monsters that feed off the flesh or blood or brains of humans but are themselves not human – any longer. At some point they were attacked and changed by one of the floating, lurching, or snarling hominids of lore. Now the undead have become a real problem or intriguing phenomenon that has seen soaring academic interest across many fields (e.g. Comaroff and Comaroff, 2002; Bishop, 2008; Fay, 2008; Davies, 2010; Waller, 2010). They have bitten into IR scholarship via Daniel Drezner's (2011) *Theories of International Politics and Zombies*, a remarkable offering by Princeton University Press. Written tongue-in-cheek, *Zombies* relates the threat of attacks by the undead, detailed gorily in novels, television shows, and films, to a range of concerns in IR. Drezner links the zombie craze to the surprise attacks on the World Trade Center in 2001 and to subsequent security measures that fail to calm nerves and actually heighten popular apprehensions and stranger-danger fears (Ahmed, 2000) associated with the era of global mobility. Vampires are sexier, although they do not reproduce sexually; zombies, however, are the ones, Drezner claims, that pose the greater risk to the public. For one, they seem plausible and have been recognized as a threat in Haitian law. The lore around zombies also has it that should they launch a war, humanity would be eradicated. And they eat human brains, the ultimate security/injury concern, one that harkens back to the Cold War panic about brainwashing techniques, nuclear attacks, and the body-altering consequences of radioactivity. Fear redoubled: surviving zombies would mutate but not die from high doses of radioactivity (Drezner, 2011: 14).

The zombie literature suggests that societies are susceptible to odd beliefs about bodies and odd notions of causality. At the same time, the monster mania provides one route of escape from hackneyed naturalist discourses that fix bodily forms, biological sex, sexuality, reproduction, and socially constructed gender as binary (even though most of the undead featured in the films and novels Drezner cites do have men's bodies). Zombies defy death, dreams of the medical perfectibility of bodies and the politics that follows from these discourses. Haraway could see that distant horizon of implications years ago. What she looked forward to, and worried about, was not the bad zombie times of our imagination. Her monster cyborgs were us, if we could recognize ourselves. They were late modern entities of inclusiveness and hybridity, rather than things set up to defeat or to eat the human.[2] Perhaps her emphasis on the machine aspect of Western ontology took her in the direction of concern-cum-celebration, whereas the post-9/11 zombie is not fired by technology but by ancient, subjugated mysteries about life and death and what can happen in the process of crossing from one to the other. Haraway suggests that beings lingering at the human–machine boundary do not fit any received story of the human, the animal or the monster and its salvation or demise. The result is that "there is much room for radical political people to contest for the meanings of the breached boundary" (Haraway, 1990: 193).

What to Drezner is a humorous development in popular culture, which shows up blind spots in IR and security theories, is for Haraway an opportunity for boundary-transgressing changes that feminists can work with to promote a range of acceptable post-gender, post-race, and post-body-bound identities. She ends one of her articles: "I would rather be a cyborg than a goddess" (p. 223), which for clarity in this context might be read as "I would rather have a cyborg body than the body, stature, and experiences of a goddess." While Drezner does give credence to IR theories that could treat a zombie attack as creating opportunities for new political understandings and actions, he has almost no time or space for feminist thinking of any kind, relegating that entire canon, including the potential ally Haraway, to two asterisked comments – not even real footnotes. It is the kind of strategy that other "critical" or "innovative" thinkers also pursue at times, the idea being to gesture to feminist analysis without actually having to read it or include its insights in their own analyses (Sylvester, 2007).[3]

Butler has always concurred that there can be no universal biological essence of "sex." Like Goldstein, she maintains that there is no direct relationship between one's body and a self or any inherent proclivities of identity and practice. Butler (1993) maintains that sex and gender are the products of language-based discourses, such as those Drezner draws on in parenthetically characterizing feminist analysis, in which certain story lines about biological sex and proper gendering gain the currency of truth and can then powerfully police bodies to produce conformity. One must perform sex and gender as expected and as institutionalized in public policies on marriage, for example, or face the consequences of social sex-gender deviance. Yet the repetition of stories and performances about sex and gender masks the socially constructed nature of the body and the limitations societies put on individual agency outside the approved texts. In her important book *Undoing Gender* (2004a), Butler begins a line of thinking that she later carries into two books on new possibilities of transnational transformational politics that are not cyborg- or undead-inspired but taken from common human experiences of bodies distant and proximate to wars. Her basic concern is to probe how bodies can gain autonomy from the ideologies and discourses of sex-gender identity and practice that she now recognizes as once bound up with community survival. How can bodies survive in today's globalized time as individuals and as societies? Her answer requires figuring out how to undo lingering sex-gender fates so as to perform bodies and lives – and imagine humanity itself – differently. She says, "to live is to live a life politically, in relation to power, in relation to others, in the act of assuming responsibility for a collective future" (p. 39). And that future, Butler argues, must not be one where injury to bodies can be part of any political-institutional *modus vivendi*.

Others, however, are critical of Butler's body approach, even though she has shifted away from performativity to the more bedrock argument that bodies are constructed socially: they do not come out of the womb with predispositions and particular needs that shape how they and society develop. Martha Nussbaum (1999), for example, finds Butler's arguments about the social construction of the body thin, myopic, and almost criminally irresponsible in the way they can ignore physical, material and bodily problems that many women experience routinely around the world – chronic hunger, beatings, rapes and the like. She wants norms of humanity, norms of human rights, to exist and to guide international policies in the interests of bodies "over there" and not only, as she reads Butler in the lead-up to the Millennium, among middle-class people in affluent societies. Butler has moved on again, though, away from her once-heavy emphasis on bodies performing and undoing the sex and gender scripts presented to them by societies, to bodies that share emotions of mourning caused by war's violence. Her turns of analysis show the difference between representing bodies as having this or that set of ascribed and enacted characteristics and experiencing social events through the body. It is the difference between talking about war and experiencing it physically. Yet physical bodies, tactile bodies can still be a stumbling-block as Butler turns toward affect studies as a basis for a transformative, anti-war politics based on common emotions of mourning felt across cultural differences, an argument we return to in the next chapter.

One other promising exploration of physical bodies comes out in a text that combines movement theory and philosophy with some IR. It is Erin Manning's *Politics of Touch: Sense, Movement, Sovereignty* (2007). Manning starts out this way: "The surface of the body is a thinking, feeling surface. It is a gestural, linguistic, sensing skin that protects us while opening us toward and rendering us vulnerable to an other" (p. 9). That body is a location of politics, of touching and resisting being touched, a mode of articulation – "the medium through which touch can be negotiated" (p. 10). For Manning, the bodily senses are relational rather than properties that people possess and show or hide. And being relational, they are also political. Manning says the senses are about the many ways bodies move and interact, with touch being "the act of reaching toward, of creating space-time through the worlding that occurs when bodies move" (p. xiv). This is not the commonsense of touch. Rather, "I reach out to touch you in order to invent a relation that will, in turn, invent me" (p. xv). We move the relation through touch. It is also not straightforward – touching and inventing, relating and being political combine several edgy elements.

To Manning, the proverbial touching and eating of the first apple can be thought of as the "violent entry into the political" (p. 49). At that biblical moment two people chose an imperfect and chaotic life on earth over a perfect heavenly paradise. Their decision can be thought of as political because it "engages us toward the world and therefore toward each other" (p. 49). And that decision was also our fall, our condemnation to death, the slipping-away of what we could know perfectly for the ominous hint of what would come next. The dual legacy of that originary touch is the bold reaching toward freedom and the equally audacious slap of sin – or of states that try to fix bodies as this or as that, here or there, in order to control them. But bodies move, are on the move, and "a sensing body in movement will always circumvent a project that attempts to characterize it in the name of touch, the senses, gender, race, politics" (p. xvi). It has characteristics far in excess of all categorizing and fixing of traits and tendencies. Bodies reach toward other bodies in forbidden as well as celebrated ways, "engaging in combinations that remind us that bodies are always stranger (*umheimlich*) than they first appear" (p. xvii). These points unexpectedly link up with queer thinking and with Michael Shapiro's (2011) earlier cited work (Chapter 1) on space–body relationships involving households and war. Clearly, a moving body is an experiencing body, a deciding body, and a political body.

There is so much more that could be said about bodies as contested entities and frameworks for experience. The literatures on bodies in the natural and biological sciences, in philosophy, and continuing in feminist analyses, could comprise whole libraries. For the purposes of this study, however, it is not necessary to range across all ideas on the body to make the central point that even though human and other bodies are everyday subjects and objects, questions outnumber answers about body forms, origins, proclivities, representations, and capabilities. In studying war as experiences of physical and other types, therefore, the body looms as a rich and promising area for additional empirical and philosophical exploration.

Feminists consider war and physical injury: Cases

It is useful at this juncture to examine how some of these issues play out and are resolved when feminist scholars investigate bodies and war. Two works are highlighted below, both of which are about bodily assaults on women in wartime. Dubravka Zarkov (2007) has studied media narratives of gendered bodies ethnicized and injured by rape in the Balkan wars that followed the break-up of the Soviet Union. Maria Eriksson Baaz and Maria Stern (2010) have done ethnographic research on the nature of wartime rape in the recent conflicts of the Democratic Republic of Congo (DRC). It is noteworthy that both studies locate bodies at an intersection of materiality, gender, ethnicity, symbols, space, movement, time and culture. As well, the rapes that occurred in both locations of war forced comminglings of bodily fluids that often left victims at the edges of life and death; the social undeaths they were made to suffer were just as injurious. And yet those sensing bodies were and are also in movement in ways that defy easy characterizations.

Bodies of Balkan wars: Dubravka Zarkov

Zarkov investigates the break-up of Yugoslavia from the beginnings of conflict in the late 1980s to the Dayton Agreement of 2005. She presents it as a case of purposeful efforts by political and media leaders to produce gendered ethnicity as a way of firing up regional hostilities. Women's bodies were politicized through discursive and physical modalities that not only led up to collective violence but also incited attacks on the bodies of women and men across the land as justifiable ways of warring. Hers is a study of narratives about women and war gathered from regional media sources.

Starting in 1987, a sizable number of women took to the streets in various locations of the erstwhile Yugoslavia to protest remarks by political leaders to the effect that women were prostitutes. The press and various feminist groups initially presented the protesting women as aggrieved mothers, an identity that many of the activist women themselves claimed. Over time, however, women's image altered in the media to the point where they were depicted contemptuously as body parts ("breasts") and as women who sold their bodies to men (whores). Ethnicity was layered onto those increasingly negative designations until the women's bodies, questionable characters, and ascribed ethnicities were made to merge. Although the media led the charge against women, local feminists were themselves inconsistent in their reception to the women's political actions. They initially criticized the protestors as women who had been manipulated by ethnic nationalism, a claim that reminds one of the feminist politics surrounding women combatants in Sri Lankan and Northern Irish conflicts. Later, local feminists changed their assessment of women's efforts to influence Yugoslav politics, praising the actions and calling it agency. Echoing aspects that Butler also discusses, Zarkov draws attention to "the symbolic capacity of the maternal body to acquire different meanings and represent different realities without any apparent contradiction . . . in the conflict zones of then Yugoslavia" (p. 70). The contradictory social characterizations of women's bodies were one sign of how some people experienced the run-up to the region's wars: via verbal and philosophical denigrations.

When the wars broke out in 1991, rumors were rampant that rape was a war strategy used especially by Serb men against Bosnian women. The Serbian press vehemently denied this and reported that Serb women were being raped by Bosnian men who hoped to impregnate them and "water down" their ethnicity. Termed the "crime above all crimes" by the Serbian press, the consequences of rape worsened for pregnant women as their delivery date approached: they were depicted as carrying "alien" children, a charge that led many raped women to consider killing their own bodies and thereby the body of the nearly mature fetus. Yet remarkably, it was the Serb man who emerged as the real victim of the rape – "'wronged' by seeing his ethnic stock polluted and his ethnic lineage disrupted by Muslim men . . ." (p. 123). As for raped Bosnian women, the press described them as ashamed and dehumanized, alive but also dead humans, and used words like "beasts" and "monsters" to describe Serb rapists. It became commonplace to equate these ethnicized women with Islam: in academic studies and in the world press, they became "Bosnian Muslims." Says Zarkov: "For those familiar with life in Bosnia the equation of the former Yugoslav republic with the traditional Muslim countryside ruled by religious customs strikes a false note" (p. 146). Over time, the Bosnian-Muslim women became identified in the international press as "the" authentic rape victims of the conflicts. They were the ones most in need of assistance in voicing their tragedies, even though many among them were outspoken about what had happened, and even though women from Bosnia were not the only ones raped.

In the Balkans, it seems that the war experiences of many women were reduced by reporters to the physical assaults of rape and to the social aftermaths of the assaults. And then rape was weighted according to the ethnicity and religion of the victim, with Bosnian-Muslim women taking the "prize" as supposedly the most frequently raped and the ones the worst off afterwards. As Zarkov puts this, "[I]t is the ultimate victory of the ethnic war and the media war that raped women and rapists were so consistently counted, included, and excluded, exclusively through their ethnicity, and because of it, that the analysis of rape repeated – instead of subverting – the ethnicization of both the victim and the perpetrator" (p. 154). Rape was written on the bodies of many women, and experienced by many as war – regardless of how women might have thought of themselves and no matter what else they did and experienced as war.[4]

Bodily violations were recorded among men in the region, too, as the final report of the UN Commission of Experts indicates. In scenes that would recur ten years or so later in Abu Ghraib, men taken prisoner often "were beaten across the genitals, forced to strip, raped and assaulted with foreign objects, and castrated" (p. 155). These were not one-off actions but politically motivated practices. Again, it is said – by witnesses, not by any male victims – that the perpetrators were Serb men and those they assaulted were Muslims. There was not as complete a silence enveloping women combatants, but the actual, rather than romanticized, experiences of those women were barely touched on in the press. Just as the castration of some men during the war was a difficult realm both of reportage and testimony, so also were the actual killing skills and activities of women soldiers. The press took an interest in them but the feminist community hunkered down around the issue of raped woman and not the issue of the women with military agency. Thereby, feminists delegitimized women's experiences of war labor and emphasized instead their victimhood, a trend in much traditional feminist analysis, as noted previously.

We learn from Zarkov's study that women's bodily experiences in wars can be, and often are, carriers of war's basic characteristic of injury. Men are injured in bodily ways in war too, but that is expected and therefore, in the context of the Balkan wars, is not commented on as much as injury to women's bodies. Women are the ones who experience the downsides of collective violence for political ends, and little else. It is an old line of thinking among feminists and general publics in the West, but the nuance to underscore here is that war as bodily experience places women front and center and not at war's margins. Another way of saying this is that in the new wars of our time, women's bodies can be required parts of direct war experience. War is conducted by and in women's bodies and men's bodies: it is just the sense of wars' large and heroic/tragic scale that keeps us from appreciating this point.

At the same time, it is important to remind ourselves that war, like other social institutions, opens doors for some and closes them for others. War can give women opportunities to change their usual labor, as Jean Bethke Elshtain (1987) put the matter of women's active participation in wars and war industries, in ways that empower them and enhance their citizenship rights. Women can fight in wars and thus injure rather than care-take others (Sjoberg and Gentry (2006)). They can jockey for leadership positions in war industries; or become advocates for women's rights in local communities and eventually win international acclaim as peace negotiators. Yet when some women change their labor through war work, they can also experience emotional injuries post-conflict, and be ostracized by their communities or misunderstood by international agencies. Moreover, heroes, victims, and villains can shape-shift over the course of a war or post-war, or blur into instances of moral ambiguity (Elshtain, 2005). That is a key point that comes up in the studies of gender-based violence that Maria Eriksson Baaz and Maria Stern have conducted in the Democratic Republic of Congo (DRC).

Soldiers talk about rape: Maria Eriksson Baaz and Maria Stern

In a number of publications, Eriksson Baaz and Stern (2010, 2009, 2008) present and analyze the results of their project to understand the complexity of rape in a country regularly referred to as the rape capital of the world – the DRC. Their study is unusual within feminist IR in going to the perpetrators rather than the victims to ask: why rape? Is rape a weapon of war in the DRC sanctioned by the military commanders or does it have other sources? Stated differently, at a time when rape in wartime has finally been recognized as a war crime in the International Criminal Tribunals of the former Yugoslavia and Rwanda, why does rape take a leap in frequency in the Congo? And why did those rapes continue after the signing of a peace accord in July 2003, when the various armed groups were being integrated into one national armed force? Some sources of military grievances are known. The integration and power-sharing schemes did not work, and combatants remained loyal to their original commanders. There was also no weeding-out of human rights violators before combining the militaries, little money for the salaries of the new national force, and little respect for the forces by civilians those same soldiers had victimized in earlier crimes of opportunistic aggrandizement. Disorderly endings of "new wars" are not uncommon, but then again, as Eriksson Baaz and Stern point out, not all such wars feature rape.

Since 2005, the two scholars have been interviewing military personnel in local languages and have now talked to more than 200 soldiers and officers, 20 percent of whom say they were ex-child soldiers. They find that virtually all the soldiers interviewed make a link between sexual violence and aspects of masculinity, and that they tend to differentiate rapes as those that are acceptable and those that are evil but understandable situationally. Acceptable rapes are committed "simply" to satisfy sexual urges. "One male sergeant likened being in battle to being in a 'desert'; the thirst of male sexuality combined with the absence of women renders soldiers parched and wanting" (Eriksson Baaz and Stern, 2009: 506). Some went so far as to suggest that one of the prime benefits of having women in the military is their presumed sexual availability to male soldiers. Soldierly masculinity, although touted by the interviewees, seemed dependent on a sexual femininity: without women present among them, the men could rape to satisfy themselves and thereby continue on with their soldierly duties. There were also those who maintained that low and unreliable pay undermined men's ability to be good family providers, which put off their wives. Rape for sex under such conditions seemed generally acceptable to them, although most noted that all rapes were morally wrong. Evil rapes were something else, a category of physical brutality and inhuman acts that left the rapists feeling shame and dishonor. Such rapes entailed not only sexual assault, but also mutilation of the victim's body, often leading to her death, as a way of humiliating and destroying the victim and her family. Said to be about "the craziness of war," evil rapes were also fed by drug-taking, hunger, and low social status, and seemed to suggest that war begets the kinds of violence that would be unthinkable and highly criminal during peacetime.

The researchers argue from their interview data that rape is not a weapon of war per se in the DRC, as it seems to have been in the Balkan wars (Zarkov, 2007; also Skjelsbaek, 2012). Rather, rape is tied up with local issues related to civilian contempt for the soldiers, mixing military units hostile to one another into one unit, militarized masculinity as a norm, plus "imagined (and real) marginalization and (imagined) needs of reasserting power and authority" (Eriksson Baaz and Stern, 2010: 57). There were no orders from commanders to rape as a first, last, or any resort. Of the three types of militarized rape that Enloe (2000) has presented – recreational, national security rape, and systematic mass rape – Eriksson Baaz and Stern (2009: 514) find that rape in the DRC has been largely recreational or vengeful. It is the result of a mismatch between the soldiers' "embodied experiences and their aspirations to inhabit ... impossible subject positions ... as "Men." ... [R]ape here serves as a performative act that functions to reconstitute their masculinity – yet simultaneously symbolizes their ultimate failure to do so." It is a conclusion in line with Elizabeth Jean Wood's (2011) analysis of rape across a number of war cases: rape in wartime has a local context that many feminist-inspired studies fail to note, leading to a general neglect of wars in which rapes and other acts of sexual violence do *not* occur, or occur with less frequency and intensity than in other wars.

Concluding observations on bodies and war

Although these are only two case studies on rape and contemporary war and cannot be generalized to other conflicts (Gerecke, 2010), they represent the types of close-up research being carried out under feminist rubrics. Such studies do not suggest that war opens doors of opportunity for many women; quite the contrary often occurs. Kimberly Hutchings (2001) warns of instantiating the moral value of war for women in feminist IR war studies, a point that this research supports in revealing the devastating bodily consequences that war can have for women and men. Hutchings also reminds us that experience in war is mediated by governmental and nongovernmental agencies. Recall that MacKenzie (2013) found international programs, such as UNICEF's Girls Left Behind, insisting that all the women who participated in the Sierra Leone war were camp followers, abductees, and victims of sexual assaults; or in the case of girl soldiers, they were "unaccompanied children," which meant they could not receive resources earmarked for child soldiers.

Eriksson Baaz and Stern have observed a parallel gap in the DRC between the good intentions of UN and Swedish agencies in emphasizing counseling for raped women, and the types of assistance many raped women there say they actually need, such as financial support for their children.[5] Anthropologist Mats Utas (2005) also talks about the global media frenzy over child soldiers that can result in children who were not child soldiers claiming that identity as their ticket out of a post-war zone.[6] Eriksson Baaz and Stern (2010), although clearly concerned about rape, also believe that too much overall attention has been paid to it as a type of war experience that is exceptionally violent. In wars, violence and injuries of all kinds are rampant and yet are implicitly treated as more normal and less troubling than rape. Why is this?[7] They note that one result of the attention directed toward war rape has been the commercialization of rape in the Congo. Zarkov talks about other consequences of media attention to rape, including the power to define ethnicity as *the* problem of Yugoslavia's break-up, in part by placing too much responsibility on raped women, who bear the new ethnic burden with their bodies.

These various warnings are important. When war is seen as bodily and emotional experiences in horrifically violent settings of politics and their aftermaths and precursors, not only does an IR-neglected range of agentic and victimizing participants, situations, opportunities, and constraints come to light, but also a range of actions of injuring and being injured emerge, as do switchover performances of moral reasoning. And there can be significant fluidity of experiences, such that one is a victim at one moment and a villain another; in the case of the DRC, some soldiers see themselves as simultaneous victims and villains. It would behoove us, therefore, to think hard about the fluidities and the bodies that have those multiple and contradictory experiences of war. Manning (2007) insists that sense does not pre-exist experience, and politics and authority are part of the mythical experience influencing our senses. Go back to that biblical apple that she talks about: "the apple is about experience, even if this experience is about the 'baseness' of humanity" (p. 50), and about a violent political decision and beginning of life outside paradise. Having reached toward one another and decided to eat the forbidden fruit, humanity hits the ground hard and violently, with the thud of finiteness instead of infiniteness, and with death ahead instead of body business forever. It is about the body becoming everyday, earthbound, and as ordinary in a way as an apple.

Move the body forward in time and the body-experience question becomes: Can we touch without incurring violent experiences that ape the proverbial biblical fall? That violence exists in the reaching toward the unknowable terrain of "God" by taking that apple in spite of warnings. Manning, herself an artist, suggests that if Michelangelo had connected the finger of God with the finger of Adam on his great Sistine Chapel fresco, unknowability could have been breached – by God, who would get a sense of what being human really feels like. That moving of the relation is not achieved. Hierarchy and unknowability thereby persist, as does a nonconnection that enables the continuing violence of original sin. At the same time, the reaching without connection maintains the human self and keeps it from being subsumed into God's sameness. That is to say, the violence of the original sin of touch is what keeps us embodied rather than operating as spirits who cannot be killed. Our bodies remain alive to touch, to spatializing, and to reciprocity with other bodies. Bodies can know the war touch as a reach that violently transforms the spaces between one another. We can also reach out our bodies to evade the Manichean wars that state sovereignty – secular God – demands that we experience. It can go both ways.

And it can go in other ways, too. Jewell Gomez's (1991) vampiric protagonist of *The Gilda Stories* is a black African-American lesbian of slave ancestry. She lives in a multicultural "family" of vampires that counts among its members a number of queer Native Americans and a couple of white men lovers. Instead of reveling in the ritual violence of bloody penetration, this family is against killing as a way of replenishing its member's bodily liquids. As compensation to the victims, the family members evolve feeding rituals that require that they enter the mind of the victim and calm his or her fears, replacing those with pleasant reveries and plans for future love and longed-for activities. This happens before the feed, which means that the vampire takes in and shares the calm that the victim suddenly feels instead of experiencing violent bloodletting. In a sense, mortal dread is replaced by something good and aspirational. We might think of it as infusing eternal life. Or it could be seen as a version of life before the fall, before original sin embodied humans as people who, to think in Manning's terms, do not always realize the power potential contained in reaching toward one another. In either interpretation, bodily blood is mixed and some measure of immortality is happily achieved in unconventional vampire style.

That blood is not the blood won from warring. It is a dream experience that resonates with Utas's (2005) notion of tactic agency, which he observed in war zones in Liberia and Sierra Leone. Rather than keeping a sense of distance from the wars' cruelties, he found that some people inventively laid claim to cruel identities as one way of coping in war and moving themselves productively into a post-conflict era of peace. Recall Shane Brighton's (2011: 104) suggestion from within IR that war disorders order in unpredictably and oftentimes generative ways. In his words, war "confronts those who experience it with the need to create – and contest – its meaning in ways that do not terminate with cessation of physical violence." Although Brighton's remarks on this are too brief, the type of philosophical-ethical abstract framework for experience that he advocates for studying war could combine with Utas's ethnographically based empirical research in war-peace zones, and with Manning's advocacy of body movement, to good effect. For one, it would encourage studying up from people *and* down from abstract frameworks simultaneously, something that neither IR nor feminism has yet been able to achieve. For another, it would remove the urge to emphasize the injuries to bodies and minds that war unleashes as experience, and also focus on what Utas (2005: 426) calls the "constantly adjusting tactics in response to the social and economic opportunities and constraints that emerge unexpectedly and ambiguously within war zones." And possibly outside immediate war zones, too, in anticipation, as Brighton (2011: 101) says, "little in social and political life goes untouched by war."

The chapter that follows turns the spotlight on emotional and socio-psychological aspects of bodily experience and war: its mixed emotions. Such emotions are not likely to be of the sort one remembers from the Rolling Stones song of the same name, which is full of enraged ambivalence as well as bodies that want to dance that rage away – although that is one possibility. Rather, it is a realm full of disorienting and generative bodily feelings and political possibilities. These can be drawn out from Judith Butler's most recent work in call-response with Erin Manning's ideas on reach and movement.

4

WAR AS EMOTIONAL EXPERIENCE

If we indeed turn the IR study of war around to include war as experience, then a discussion of bodies must expand beyond the physical to the realm of emotional experience. In some ways, the body-mind-emotion division is an awkward separation; in the view of many, emotions are part of bodies. What the body experiences, such as rape or another type of war injury, produces not only physical responses but emotional responses as well. A person can turn fear, anxiety, exhilaration, confidence, insecurity, anguish, and love into a host of other emotions in situations of war. That is obvious. What those emotions are and what they indicate, as well as how they are constituted and how and whether they become political, are less obvious or easy to study. Feelings and emotions operate on a terrain of tremendous controversy, and again those reigning debates take place mostly away from the portals of IR. The rationality assumption, which black-boxes emotions or considers them dangerous flaws in decision-making, remains one of the field's most enduring credos, no matter that neuroscientic evidence suggests reason and emotion operate in tandem. The critical wings of IR take up the subject of fear in talking about the climate of national and international relations post-9/11. But is fear an emotion or an affect? And where does it reside – in bodies or in culture and society? Feminists are not afraid of that question; indeed, they are comfortable with a world of feelings, emotions and reason. They do, however, disagree among themselves on where emotions lodge and what they mean.

It is possible that external influences trigger emotions, something akin to marketing techniques stimulating demand for a product in consumer economies. Susan Bordo (2003) expresses concern that young Western women *are made to feel and to think* that something is wrong with their bodies – overweight, wrongly shaped, flabby, or absent celebrity. The fashion/beauty media are a big influence, as are reality television shows and cosmetic companies. Social gender norms also affect what people around the world can feel about themselves in terms of mothering, sex, manners, clothing, and aspirations. No matter how a girl might long to be a Catholic priest, the fact that her mind and her devotion are carried in the wrong body eliminates her from consideration; she is meant to feel ok about that. If she lives in a Confucian society, she is expected to be content at a relatively low status in the social hierarchy. If she is a Muslim woman living in certain communities in the Netherlands, she might, as Hirsi Ali (2006, 2007) has noted, be confined to the house unless accompanied by a male, and be expected to accept that. If s/he is in the military, basic training programs will inculcate feelings that not keeping up on the morning run or other physical challenges are signs that one is a weak girl rather than a properly militarized "man."

If a person does not feel content in an externally produced status, body punishments are at hand, such as whippings, kitchen "accidents," stonings, ostracism, mental institutions, torture. The tighter the norms of social and political correctness, the sharper the punishment for transgression, along a line that starts with feeling guilty, feeling angry and feeling very threatened and punishment can, as Karin Fierke (2004) suggests, lead to feeling numb or vengeful on a scale that engages entire societies. Betty Friedan (1963) talked decades ago about the problem that had no name: galloping unhappiness among middle-class, pre-feminist era women who were stuck in low-paying jobs or meant to feel fulfilled at home in American suburbia. Because their gender problem had no name, a woman could easily think that something was wrong with her alone – she was maladjusted at minimum and possibly mad – rather than noticing that something was wrong with gendered society at large. More recently, Sara Ahmed (2004: 119) argues in a parallel vein that emotions are not private feelings that come from within oneself and one's particular body. Emotions move between bodies, "align individuals with communities – or bodily space with social space – through the very intensity of their attachments." Emotions produce (and sustain) the ordinary, that community of time and place that Veena Das (2007) urges researchers to enter into rather than rise above.

In this chapter, we expand the discussion of bodies from Chapter 3 to take into account views on emotion as body-based relative to emotions as the political economy of external influence and manipulation. The initial focus is on an exemplary text on IR and emotions by Neta Crawford (2000), where a wholly workable definition of emotion is offered, and also ideas that percolate through a more recent Forum on Emotion and the Feminist IR Researcher (Sylvester, 2011b). The ways in which these IR writings explore emotion are set against the backdrop of work in the neurosciences and neuropolitics, in which affect is separated from emotion (Massumi, 2002) or is linked in a complex network of relays that William Connolly (2002) articulates. Issues of emotions and war are then taken up, with Judith Butler (2004b, 2010) in the lead and Erin Manning (2007) in implicit dialogue with her on concerns about mourning and anti-war politics. A final section on war's enthusiasms pulls us up to the realization again that war is about injuring and injury, but that there are some for whom combat is also a place of unexpected well-being. The exercise of comparing and contrasting approaches to emotion keeps enlarging and nuancing the ongoing discussion of war as experience. War engages physical bodies in all the complexities noted earlier, as well as in the emotional reactions that attach to bodies, even if some emotions are not authentic or original to those bodies; after all, gender is not "real," and that point has had no hindering effect on feminist researchers who explore "its" importance in society and the many ways gender is experienced, shaped, and influential in real lives and real socio-political institutions.

IR considers emotions

It is not fashionable outside many feminist traditions to link emotions directly to the body, to see emotion, that is, as coming from within rather than traveling on the connective surfaces of bodies, as Ahmed (2004) argues. Much as IR focuses not on war per se but on its abstract causes, the emphasis in much social science writing is on emotion-constituting influences and their effects on the body politic (minus the bodies). Causes, sources, locations, forms, and measures are important, but emotional experiences that draw on the bodily senses – not remaining there necessarily, but starting there – get less attention. We can see that tendency in recent key works in IR that show unwillingness to look at all emotions and international relations, let alone at emotions and war. Yet there is one key text that carefully navigates the debates and ends up proposing a definition of emotion that is adopted here.

Passions and politics: Neta Crawford

Neta Crawford's (2000) "The Passion of World Politics" is the most significant recent piece on emotions in IR. It argues that IR harbors assumptions about emotions that it does not subject to examination. The longstanding belief that actors are rational, in the sense that they weigh options and make logical cost–benefit calculations of gain and loss, blinds the field to the influence that emotion has on rationality. In the realist foreign policy analyses of John Mearsheimer and Stephen Walt discussed in Chapter 1, for instance, it is noteworthy that America's unconditional support for Israel is characterized as not rational, not interest-based and therefore as dangerous. What the support reflects is a persistent compliance with emotion- and power-laden insistences of a major lobby group. Beyond rationality lies the territory of bias, misperception, and erroneous decision-making – indirect terms all for the messy and faulty realm of cognition, affect and emotion that can only yield up mistakes.[1] And yet, as Crawford is able to show, realism can be full of emotional influences like fear – just as its scholarship can harbor people who are fearful, as I showed in Chapter 1.

A dichotomous view of how emotion operates afflicts studies of war as well as conventional-security, strategic, liberal, realist, and rational-choice IR theories (e.g., Walt, 1999, Fearon, 1995). American approaches to IR can game war, work out optimal and suboptimal decisional options through formal modeling, and discuss decision-making in terms of what Crawford calls psychology's once dominant emphasis on "'cold' cognitive processes" (p. 118). By contrast, the master European war theorist, Clausewitz, whom Crawford quotes (p. 121), said war was composed of "primordial violence, hatred, and enmity." More contemporary European analysts, or those strongly influenced by the twentieth-century wars of that continent, remain less inclined to write off emotions in/and war (e.g., Wright, 1965; Morgenthau, 1965; Lasswell, 1965). And of late, constructivism has reawakened American IR to the possibilities of emotion in the international through its emphasis on norms (Finnemore and Sikkink, 1998; Keohane, 1990), while scholars following European traditions have added to research on emotion and international relations by studying aspects of trauma (Edkins, 2003; Fierke, 2004) and specific emotions of pity (Aradau, 2004), empathy (Sylvester, 1994b), forgiveness (Gobodo-Madikizela, 2008), and relations of fear and security (Huysmans, 1998; Kinnvall and Linden, 2010). Running through these works are concerns that emotion is poorly understood in and amongst a range of mental states often randomly called feelings, desires, affects, perceptions, and cognitions. And then there are methodological issues: "valid measures of emotions are not obvious; and it may be difficult to distinguish 'genuine' emotions from their instrumental display" and their cultural surroundings (Crawford, 2000: 118).

For Crawford, the leading problem in this area for IR is that emotion has not been seen as consistently salient and therefore has not been properly defined and elaborated. She offers a number of propositions about emotions that she urges IR to flesh out. These include the expectation that emotions are ubiquitous but will vary in expression, intensity, and behavioral manifestation, and the idea that emotions will "influence the performance and content of information gathering and processing by individuals and groups" (p. 137). To her credit, she does not encourage competition between those who see emotion as body-based and those who see it as mostly socially constructed. Her concerns are far broader than that: to encourage everyone in IR to consider the usefulness of propositions on emotion for their research on "foreign policy decisionmaking, war, peace, and diplomacy" (p. 130). The propositions themselves are carefully articulated and inclusive of experiences that range across the many terms that are often used interchangeably with emotions. Crawford gives very short shrift, however, to feminist IR contributions to theorizing the significance of emotion to international relations, offering only two citations in that area and virtually no gender examples that relate to international relations. She also strays little from the established sense that emotion has importance mostly for studying prominent decision-makers and decisions, and not for studying people on the ground who bear the burdens of those decisions.

Crawford's working definition of emotions, however, is the one I find most useful for studying the emotional aspects of war experience: "the *inner states* that individuals *describe to others as feelings*, and those feelings may be *associated with biological, cognitive, and behavioral states and changes*." She continues on to say that "[f]eelings are internally experienced, but the meaning attached to those feelings, the behaviors associated with them, and the recognition of emotions in others are cognitively and culturally construed and constructed" (p. 25 emphasis in original).

It is a definition that locates emotions in the body and yet recognizes that their interpretation will reflect the social relations in which the body is embedded.

Although there has been an upsurge in interest in emotions and international relations since her publication, owing to what Crawford could see as an unfolding emotion revolution in psychology, recent IR work still assumes that if emotions are studied at all, they will be the emotions of state or elite decision-makers. It does not descend long into the realm of ordinary people and their emotional experiences of international relations. Such people enter briefly, if at all, to illustrate abstract points about how emotions work or should be studied (Hill, 2003; Balzacq and Jervis, 2004; Lebow, 2005; Mercer, 2005, 2010; Bleiker and Hutchison, 2008). And so far, no discussion of the status of emotion in studies of international relations has given substantive attention, as opposed to simple citational attention (and usually no attention at all) to the emotion-related research conducted under feminist IR rubrics (e.g., Cohn, 1987; Tickner, 1988; Sylvester, 1994b, 2000; Stern, 2006; D'Costa, 2006; Jacoby, 2006).

Forum on emotion and the feminist IR researcher

These IR orientations are turned around in a group of six essays that appear as a forum on "Emotion and the Feminist IR Researcher" in *International Studies Review* (Sylvester, 2011b). Their starting point is the researcher experiencing her research and the profession of IR as a point of usual invisibility in conventional IR studies. Each contributor to the forum endeavors to figure out why the researcher goes missing in her own feminist IR writings, even though feminist theory clearly provides for her in its equation of the personal with the political and the methodological admonition to establish researcher positionality (see Chapter 2). This is a particularly thorny issue when the research is ethnographic in approach and the research site is a recent war zone. What does the researcher do with her fear, excitement, sadness, pity, alarm, or anger in the field? Pretend it did not exist when writing up the research? Act as though the traumas of difficult fieldwork just pass once the researcher leaves the scene?[2] What about the possibility that researcher emotions can affect the emotions of people being interviewed and thus their responses to questions? What about personal emotion clouding research judgment? How do the accounts people give of the injuries and exhilarations of war around them affect the researcher's writing and logic? Where in the conventions of social science writing can one detail the emotional side of the research? These are decidedly not questions that usually emerge around emotions and international relations. The struggle for answers says much about how little we probe the relationship between the research process and the words that form on our pages and rack up, or not, as research results.

The opening essay of the forum features an advanced women's studies MA student expressing amazement at the straightforward manner in which feminist IR war researchers discussed difficult fieldwork experiences in war and conflict zones at a workshop on women and armed combat.[3] Sandra Marshall, its author, notes the unwavering composure as they spoke of their research encounters; they seemed unaffected. I was steeped in the research ethics of such feminists as Luce Irigaray (1985), Ann Oakley (1998), and Donna Haraway (1991)—to name just a few of the theorists who have called for researchers to include themselves and their emotions in their research. My feminist alarm bells therefore started ringing ... Were these feminist IR researchers genuinely unaffected, or were they just keeping quiet? (Marshall, 2011: 3)

She endeavored to find out. She interviewed several workshop participants about their experiences and also examined other feminist IR works to see whether researcher emotions were discussed in publications. Marshall found that feminist IR literature emphasized the emotions of subjects studied and not at all the researcher's emotions; she sighted only hints of personal trauma and concern as "hauntings" in the texts. That area of silence ran in the opposite direction to feminist discussions developed outside IR, where some inclusion of researcher presence and experience was encouraged as methodologically sound and ethically necessary.

The scholars she interviewed told her, and also wrote in the forum, about situations they confronted on the ground that produced personal responses they were ill-prepared to experience and uncertain how to process. Operating in a social science universe that continues to reward "objectivity," they were initially silent about their experiences doing war research, even though Megan MacKenzie (2011b: 5) acknowledges that "[f]rom the minute I landed in Freetown until the day I departed, emotions completely hijacked my original, concise and 'rational' research plan." She says that "[i]nstead of writing about the experience of my field research and how it impacted my overall approach to IR, I consistently wrote this *out* of my work" (p. 5, emphasis in original). Through the experience of being interviewed by Marshall, MacKenzie came to the realization that writing experience and emotions out of research "is at the heart of what I see as the 'problem' with emotions and IR ... it is necessary to encourage scholars to write themselves back into their scholarship" (p. 5). To do so is not to indulge in personal biography, seek attention, or privatize the research experience. Rather, MacKenzie expresses concern that the absence of interest in researcher emotion in "data"-gathering contexts constitutes, as I interpret her words, a research problem without a name that impacts on the quality of the studies produced.

Swati Parashar (2011: 9), who has interviewed militant women in Kashmir and Sri Lanka, also recognizes the ways she hid within her write-ups of their efforts. She argues for acknowledging that researchers are "constructed by desires, emotions, interests and politics while doing research. Knowing is intrinsically related to feeling; and yet we are unwilling to take intellectual risks by researching/acknowledging/ writing the emotional into our works ..." She very specifically draws attention to feelings of alienation from the subjects she studied and to decisions she often made "about which identity I could privilege for maximum benefit to my research and minimum conflict with my subject. Sometimes, I would willingly adopt the anti-India position to be on common ground with my interviewees. All these multiple performances did affect my nerves at times ..." (p. 11). In other words, Parashar engaged in masquerades of research practice instead of leading "with the idea of reciprocal subjecthood" (p. 11); and those masquerades affected her work.

Although the forum focuses on the missing researcher in the research, especially in situations where the researcher is operating in challenging conflict zones, it is very characteristic of feminist IR scholarship to take up issues like this that the wider field of IR neglects or misconstrues. Parashar: "If as feminists, we privilege the 'personal' as 'political' and as international, the 'personal' must also be (re)presented through a ubiquitous set of emotions that govern actions, behavior, and responses" (p. 11). As several analysts bemoan, IR has not taken emotion and politics to heart as a continuous problematic in the field. Interest spikes now and then, as during the Cold War years of the 1950s and 1960s, only to be followed by long troughs of neglect. The feminist IR forum is novel both for its insistence that an IR enamored with nearly emotionless rationality must keep emotion in its portfolio, and its equal insistence that there is more *to* the study of emotion than "faulty" elite affect and cognition. Emotion is ubiquitous, as Parashar states, which means that it exists in the researcher, in all subjects, and in interactions between researcher and subject(s).

There is debate in feminist IR circles, in the forum, and in feminism more generally, however, over exactly where emotion is located, produced, and experienced. Within the forum, Shirin Saeidi and Heather Turcotte (2011) contend that personal emotions are conditioned by power structures and historical legacies rather than individual experiences. The feminist recognition that the personal is political cannot, they say, be allowed to be a position in which the "*personal is privatized*: We understand the personal is privatized as a political struggle that silences systemic inequities and limits access to the claims of personal politics; it individualizes the *moment of exchange* and disconnects it from larger circulations of power" (p. 8 emphasis in original). Focusing on individual emotion deflects attention from issues of inequality that could foreground "larger structures of power that connect us to one another, especially within the discipline" (p. 8). Their concern is to shift the level of analysis from individuals to power relations that circulate historically and through contemporary institutions in ways that shape feelings about self and other.

That argument resonates, as Saeidi and Turcotte acknowledge, with a body of work on emotion and politics exemplified by Ahmed's studies on emotion as an affective economy. To Ahmed (2004) emotions circulate between bodies and signs, and are categorically not private in origins or in their places of residence. Emotions sit on the surface and flitter back and forth between bodies to create communities of like feeling, such as the nation, with attendant feelings about who belongs and who is a despised "other" fantasized as causing injury and pain to the true community. To move in that direction is to conceptualize feelings as operating as economies rather than as psychological dispositions; indeed, Ahmed points out that psychoanalytic approaches show that "emotionality involves movements or associations whereby 'feelings' take us across different levels of signification, not all of which can be admitted in the present" (2004a: 120). That movement forward and backwards and sideways means, for Ahmed, that "emotions work as a form of capital" that circulates and is distributed across social as well as psychic fields (p. 120). Emotions accumulate over time and shape policies, a point that resonates with Fierke's (2004) framework for understanding war trauma (see Chapter 1). But the key point for Ahmed is this: "it is the failure of emotions to be located in a body, object, or figures that allows emotions to (re)produce or generate the effects they do" (p. 124).

Hers is a strong and provocative statement about how collective feelings are formed and become political. Saeidi and Turcotte (2011: 9) modify the strength of Ahmed's attachment to the economic model slightly by indicating that "our personal emotion is importantly informing these affective knowledge exchanges," thereby giving some recognition to the personal and implicitly to the bodies of persons experiencing emotions. They go on to say, however, that "we need to be attentive to the ways in which our centring of the personal is part of a longer academic legacy of affective and "discursive colonizations," (p. 9; also Mohanty, 2003). Feminist IR, they argue, has isolated itself from postcolonial analyses of "theories of the flesh" (p. 8), which is unfortunate, for those theories suggest that "the individualizing of struggle erases the systemic" (p. 9). Parashar objects to this framing. Having lived with and daily confronted the aftermaths of colonialism in India, she argues that Saeidi and Turcotte overstate the systemic. Do they not thereby "produce the very 'differences' and divisive power relations that postcolonial analysis decries"? (Parashar, 2011: 13) She explains:

> I am thinking of the tendency in postcolonial analysis in the West to sustain the "subaltern" as a subject status that cannot cease to exist, and which is peopled by those who cannot have "personal" emotions if they have also experienced colonization, slavery, and capitalism. That argument can be depoliticizing and disempowering ... sweeping and inadequately contextualized. As one positioned to know the colonial nature of power, I also argue for my own subjectivity and emotions as a researcher and for those of the researched ... without assuming an unchanging discursive framework of colonized spaces and affectivity that can end up iteratively reproducing differences instead of also producing differentiated agencies. (P. 13)

The debates on where emotions come from and the degree to which bodies are their originary receptacles will undoubtedly continue. Judging from the forum, feminist researchers operating in difficult war and conflict zones experience body-based reactions to the social conditions of the research. Other scholars argue that politically salient emotions are produced and cultivated by external events that stir leaders and can lead them into policy mistakes (Mearsheimer and Walt, 2006; Mercer, 2005) or into demeaning manipulations (Fanon, 1963) that could include state "security" practices that foment high levels of public insecurity post-9/11. Yet to argue that the body is mostly out of the picture of emotion except for surface flutterings that take place along bodies en route to someplace/someone/

something else can seem too much of a stretch to some, as it does to Parashar. Crawford (2000: 125, emphasis in original) wants to define emotions as "the *inner states* that individuals *describe to others as feelings* . . ." Clearly, she is lodging emotions in bodies while also acknowledging that the meanings and behaviors attached to those feelings are both cognitively and culturally construed and constructed." Parashar's view accords with this definitional preference for locating emotional experience in the body. At the same time, the Crawford definition also shows awareness that emotions have historical power contexts, as Saeidi and Turcotte note. Those power contexts are expected to change over time – again, this is Parashar's point – resulting in changes to subjectivity, interpretation, behavior and status, all of which researchers must note. This capacious definition provides room for emotions and for changing relations of agency at the bottom of any power pile as well as at its top.[4] But it is not the only way forward.

Related views of emotion

Chapter 2 introduced the concept of experience as explored by a number of feminist and other thinkers in the 1980s and 1990s. The sense that emotions can change over time requires us to revisit and add onto that discussion: experience with experience as a concept can also change. Lauren Berlant (1993: 571–2fn) suggested one direction research on experience could go when she offered the view that experience is "an activity that becomes framed as an event . . . something one 'has,' in aggregate moments of self-estrangement." In subsequent years, historians, geographers, biological and neuroscientists, political theorists, art and architecture students, literary theorists, and cultural and media studies have become fascinated with the intricacies and technicalities of feelings. New disciplines devoted to such topics have arisen – neurohistory, neuropolitics, and neuroaesthetics among them. Many of these endeavors echo Berlant's implicit concern to turn murky and encompassing concepts like emotion, cognition, and affect (which Crawford puts into the same frame) into more specified phenomena. Feminist Martha Nussbaum (2004) is one among those seeking greater precision. She advocates a cognitive approach that locates emotions as elements of intelligence. To her, emotions are appraisals or value-judgments that relate to a person's goals and capacity to flourish and that also register how we are viewing external situations that could bear upon our well-being, in general and with respect to the liberal state. Given a concern with the moral/ethical judgments that emotions express, Nussbaum proffers certain emotions as fitting life in the liberal state well – anger and indignation; fear and grief; gratitude and love; compassion – and others, such as disgust and shame, as inappropriate.

Of late, however, neuroscience and affect theory in cultural studies dominate the study of emotions. They look to the body as the experiencing and processing vessel, but render that body more dynamic and unpredictable than was once assumed, and less ruled by cognitive value-judgments. Humanities professor Ruth Leys (2011) reviews the turn to affect studies, with reference to the work of philosophers

Brian Massumi (2002) and William Connolly (2002), who, she says, insist that "we human beings are corporeal creatures imbued with subliminal affective intensities and resonances that so decisively influence or condition our political and other beliefs that we ignore those affective intensities and resonances at our peril" (Leys, 2011: 436).

Intensities and relays: Brian Massumi and William Connolly

For Massumi, affect precedes that experience-producing event of which Berlant speaks. It is an intensity that a body senses unconsciously and unintentionally, and not something it "has" at times of self-estrangement only – when "the subject enters the empire of quotation marks, anecdote, self-reflection, memory" (Berlant, 1993: 571–72f). Emotion comes a split second later as a conscious reflection, interpretation or reaction to affective intensities. Massumi (2010) offers a brilliant, perversely hilarious, and also slippery sense of what he would term affective political experiences that are orchestrated by state authorities intent on influencing the feelings of citizens. He describes George W. Bush's efforts to rally Americans behind his argument that the invasion of Iraq in 2003 was justified despite the absence of weapons of mass destruction, because the weapons were felt to be real: "Bush did what he did because Saddam [Hussein] could have done what he didn't do" (p. 54). With this affective political warning ringing across America, he won re-election. Or there is the case of an all-out toxic substance alert at Montreal airport that saw SWAT teams descend on a leaking suitcase after closing all roads to the airport, corralling the passengers, and getting helicopters circling above. The dangerous substance warranting all this securitization was white flour. Massumi: "The preemptive measures cause the disruption to the economy and everyday life that terrorist attacks are designed to produce beyond their immediate impact … [and] the incident is left carrying an affective dusting of white-powdered terror" (pp. 57–58). The "affective dusting," however, can be read as a set of commingled responses – affective in the little dreads set off and simultaneously emotional in the meaningful, political registers produced – from annoyance and inconvenience to political appreciation of how alert the security authorities are – that also constitute the war-on-terror experience.

Massumi's position has considerable support among neuroscientists and psychologists, and in turn supports assumptions about emotions underpinning most contributions to the Forum on Emotion and the Feminist IR Researcher discussed above: first one feels something unexpected and unarticulated while doing research and later identifies it and reflects on it. Both experiences feature raw affect, which Massumi (2002: 27) calls nonconscious intensity, and social relations that give affect subjective content. Neuroscientist Anthony Damasio (1994: 158; also 2000) talks about the brief half-second between perception of something and consciousness of it and response, noting that "the brain probably cannot predict … all the imponderables of a specific situation as it unfolds in real life and real time." To talk about emotions and the feminist IR researcher, emotions in international relations, or war as emotional experience is to approach a tangle of biological as well as social components that move perceptions and unarticulated bodily intensities into meaningful registers of interpretation and politics. The danger is that emotions can be conflated with a range of "visceral forces, beneath, alongside or generally other than conscious knowing ... that can serve to drive us toward movement, toward thought, and extension ..." (Gregg and Seigworth, 2010: 1).

Yet there is that nagging sense that emotional experience must reside in some commingling of affect and emotion. Leys (2011) is not convinced that the two can be separated and then rejoined only once socially engaged emotions emerge to give affect content. She sees in the "new affect" approach a replication of Cartesian body dualism, with the raw intaker of events and feelings somewhat separate from the processing mind. Implicitly, she prefers a more equivocal and open-ended sense of body–mind reciprocity, of the sort found in earlier treatments of affect, like those of William James (1976), although that point is not elaborated in her piece; nor is it clear to Connolly (2011), who replies to her article. The one element that is clear, as Leys (2011: 471) puts it, is that there is an "ongoing clash between competing ways of thinking about the emotions."[5]

Connolly's (2002) ideas on these matters present one type of clash. Often spotted giving papers at IR conferences, Connolly holds an important crossover position between poststructuralist political theory, new affect science, and cultural studies. He launches his prominent study, *Neuropolitics*, with his own experience of living with a family situation in which his father sustained and partly recovered from brain damage suffered in a car accident. Widely known as a poststructuralist thinker, the experience of watching his father regain functional but incomplete and also new capability leads Connolly years later to take a step away from poststructuralist fascination with representations. In *Neuropolitics*, Connolly considers "the *compositional* dimension of body–brain–culture relays ... the politics through which cultural life mixes into the composition of body/brain processes. And vice versa" (p. xiii, emphasis in original). He is particularly interested in how thinking operates dynamically with the corporeal body to produce new ideas and ethical judgments; thus his interest in "relays" rather than notions of "circulation" that conjure up closed economic systems. He is also not interested in defending culture-dominant approaches or encouraging the compartmentalization of cognition, affect, intensity, and the body. *Neuropolitics* is a discussion of the bodies/emotions linkage to the world via contemporary neuroscience, a field historically avoided by poststructuralist scholarship as hopelessly tied to biological reductionism.

Integral to Connolly's vision of body–brain–culture linkages is the challenge of crossing levels of analysis by going from individual perceiving bodies to collective cultural bodies of thought and activity. It is a challenge analogous to studying up to war from the people who experience it rather than siting the key elements of war in abstract causes or rational strategies. Connolly argues that culture is layered, with each layer in an interactive relationship with other layers, producing results that depend on such factors as capacity, speech, and speed. Thinking is also layered, as several cases he presents show, with common institutions playing strong roles in shaping individual and seemingly private thoughts, such as what it means to live with a person and how to do so in socially acceptable ways. Presumably, war as a social institution affects thinking and culture in similar ways, and if we insert "war" into a rephrased litany of social considerations Connolly mentions as influencing individual decisions to marry (pp. 19–20), we come up with these: it helps if a state and its formal military sanction the idea, hold ceremonies around war, and provide assurances of support for the progency in case of death; it helps if war has been represented in routine ways in TV dramas and films; if the military you seek to join is consonant with the job prospects available to you and if neighbors and family fold war into routine practices of friendship, sociality, holiday celebrations, and so on. This is one kind of relay:

> [t]he objective institutions that preexist us become infused to variable degrees into dispositions, perceptions, beliefs, and resistances we share and contest with others. If thinking helps to compose culture, the objective dimension of culture helps to compose thinking, making the relays and feedback loops that connect bodies, brains, and culture exceedingly dense. As you attend to the complex relays joining bodies, brains, and culture, the hubris invested in tight models of explanation and consummate narratives of interpretation becomes vivid. (P. 20)

Connolly's approach helps us think about micropolitical dimensions and relays that layer with macropolitical dimensions to compose emotional experiences of war. He suggests that the micropolitical realm is generally evident in "organizing attachments, consumption possibilities, work routines, faith practices, child rearing, education, investment, security, and punishment" (p. 21). These are precisely the arenas of practice that feminist IR analyst Cynthia Enloe has repeatedly dubbed militarization, the politics of which can operate, as Connolly (2002: 21) puts it, "below the threshold of political visibility inside every domain of life." Although it is below obvious detection, Connolly believes that micropolitics works through hidden and often manipulative practices that are "mobilized to organize the affective element of intersubjectivity" (p. 109). As an example, he mentions the 1980s generation of nuclear weapons designed to get around restrictions on missile launch sites—multiple, independently targetable re-entry vehicles (MIRVs)—which became politically acceptable through a micropolitics that operated anxiously around dinner-table conversations and in the media. Thinking in parallel ways, we might argue that emotional reactions to war scenes watched on TV or followed in films become some people's main war experience. Their bodies process the incoming information and generate emotions that combine the new with memories, histories, personal experiences of other sorts, and countervailing information. People can thereby be recruited to war through the visual and verbal technologies used to transmit it (Butler, 2010: xii). This way of thinking about emotion comes at the issues differently but ends up well in line with Crawford's statement of emotions as inner states that are affected by cognition and culture. That process is layered and relayed through contingent brain-body-culture connections operating at different speeds and intensities, which intermingle past, present and even uncertain future compositions.[6]

Studies of emotions focusing on what they are, where they come from, and how they operate with other cognitive and affective elements show that models of rationality, although heuristically interesting, are not scientifically substantiated as operating in the ways that social science, and much of US IR, thinks. Research on body-brain interactions indicates that emotion works with reason, not against it or apart from it. Decisions made in the absence of emotion can be problematic or criminal. Thus, the 2011 bombings of government buildings in Oslo, combined with a shooting spree afterwards, carried out in what witnesses and the perpetrator himself describe as a state of cool unemotionality, suggests that those acts could not be rational, no matter how the perpetrator tried to present them; and indeed, the courts have debated whether the perpetrator suffers from paranoid schizophrenia or not. Because IR has identified itself over the years with assumptions that states are rational actors, decision-makers operate rationally, and rational choice is the most scientific and accurate way of describing, explaining, and predicting decisions, it is exceptionally resistant to the countervailing evidence – even though much of IR believes in science. The related idea that emotions throw off reasoning and produce mistakes is equally far from the mark. So is the historical sense, seen in the writings of philosophers like Rousseau, that emotion is something that women suffer, owing to their reproductive-oriented bodies, and that men, to their "credit," can contain. Women are not the emotional sex. Emotion is ubiquitous in human behavior and variable in its manifestations.

Emotions and/in war

There are many technicalities, intricacies, and debates in neuroscience, neuropolitics and affect theory in cultural studies. I find it useful as an initial position to define emotions in the broad way that Crawford does. That is to say, it does not seem necessary to reach a "true" position, or come down on one side or the other of the argument in order to proceed. I reach this conclusion in part because I take as a guide not only the neuro approaches but also the insights of feminist and feminist IR research. Studying up is so new in IR war research that the key point might be to keep one's senses to the ground and attuned to people's war "ordinaries" (Sylvester, 2010b). That could entail approaching experience phenomenologically, as Brighton (Chapter 1) does, via the interviews in the field that Parashar, MacKenzie, Kwon and Utas have conducted, through the discourse analysis Wibben presents (Chapter 2), or through contemplations of art and literature (e.g., Gibbon, 2011; Sylvester, 2009). Whether such approaches yield "correct" understandings of emotions relative to affect, cognition, feelings, and the like seems less central to the apprehension of emotional experience in war than does the process of opening doors for the ordinary to enter into standards of knowledge and comprehension. In other words, deferring to people's sense of their own emotions around war, however "wrongly" specified these might be in scientific terms, would matter more than standards set by the academy or policy circles of various kinds. I would encourage war studies approaches that openly embrace methodological diversity while critically acknowledging the so-called incommensurabilities that this yields, as well as issues about whose knowledge is dominating, which Gayatri Spivak (1988) brought to academic attention in the 1980s.

More salient to issues of emotional experiences of war is this question: what are the emotions of war, what do they feel like, and what cultural and political activities do they produce and/or reflect? Other than narrow debates about the existence of a small number of essential emotions all humans share, there has been relatively little discussion in the social and neurosciences about what emotions actually feel like (compared, say, to affect), and how people describe them. That question is of considerable interest to novelists, artists, and memoirists who work with war and related topics. It is also of interest to feminist analysts. Lauren Berlant's (2004) collection of essays on compassion, referred to earlier, is the lead into a more encompassing discussion now on Judith Butler's sense of shared emotions of grief.

The vagaries of compassion: Lauren Berlant

Compassion is obviously one emotion among many, but Berlant introduces it in terms that have broad significance. She suggests that it is something one feels and also "a social relation between spectators and sufferers, with the emphasis on the spectator's experience of feeling compassion . . ." (p. 1). The concept of the spectator is an important one for the study of war as experience. It taps into the sense that in a globalized era war touches everyone, even those whose bodies are not in any direct line of fire: people feel the war touch directly, through news reports, aid and relief campaigns, books, films, visual art, discussions in schools and universities, through philosophical inquiry, or by knowing someone who is or has been involved in war. To my mind, there are at least three spectator degrees of overlapping separation from war as an immediate body-injuring set of practices.

People can be close to war but behind the lines as medics, military caterers, relief workers, religious figures, soldiers waiting for action or stunned by it, or local people at a short distance from active war zones, so short that war sounds echo in their ears (these people lead anxious existences or seek to leave their home areas); family members of people caught up in war or participating in it as combatants are also in this category. At a second degree of separation are those who engage with war at a farther distance, but in very active and material ways, by producing war material of all kinds, engaging in rehabilitation of war-injured people, researching, writing, or creating artistic works about war or technological enhancements of war capabilities; state leaders and politicians are also in this category, as are war protesters. At the farthest physically removed level would be people who go about their daily business with little or no interruptions or changes owing to war anywhere; these spectators see war clips on television, read about wars in news media, play at war through video games and other war-mimicking activities that might be "innocent" or entail criminal activities, or study various wars in school as part of the normal curriculum; though the farthest removed from war, people in this third category can have emotional experiences of war. Indeed, emotions, whether expressing compassion, anger, sadness, fear, happiness, or love characterize war spectatorship at each of the three spatial degrees of separation from war. People can move between spectator levels or occupy more than one location at a time – say, as a weekend-only war protester or someone in the third level of separation who learns that his friend's daughter serving in Iraq has been killed.

The sufferer is also a relevant concept and subject, of course, one that needs probing and nuance (Spelman, 1997). Physical bodies can suffer mightily in war, through direct injuries, through the need to move away from conflict rapidly but without a clear idea of where safety lies (recall Deogratias's continuous fleeing from *genocidaires*, discussed in Chapter 3). Bodies can starve in war zones, dehydrate, freeze, and fall ill from impure water or improper sanitation. Damaged buildings can collapse onto them, or they can be burned or electrocuted on fallen power lines or in gas explosions. Physical symptoms of post-traumatic stress disorder or fears – shaking, amnesia, insomnia, and broken heart syndrome (learning someone close to you has died, causing you to have a heart attack) are common. Emotions intertwine with the body and create suffering from phobias, depression, psychosomatic illnesses, haunting dreads and anxieties, flashbacks, intrusive thoughts and memories. Yet the war sufferer can also be interchangeable with the war spectator in refugee and asylum camps or in relief lines.

Chimamanda Gnozi Adichie makes these interconnections vivid in her novel, *Half of a Yellow Sun* (2007). It depicts the Biafra War of the 1960s from the perspective of people initially enraptured by the Igbo nation's audacious break from Nigeria. As events turn against Biafra – Britain intervenes militarily against the rebel area and an embargo halts food deliveries – everyone in the region becomes a simultaneous physical-emotional war sufferer and spectator:

> A small crowd from the refugee camp was beating and kicking a young man crouched on the ground, his hands placed on his head to shield some of the blows. His trousers were splattered with holes and his collar was almost ripped off but the half of a yellow sun still clung to his torn sleeve . . . The soldier had been stealing from the farm. It happened everywhere now, farms raided at night, raided of corn so tender they had not yet formed kernels and yams so young they were barely the size of a cocoyam . . . The soldier got up and dusted himself off.
>
> "Have you come from the front?" Kainene asks.
>
> He nodded. He looked about eighteen. There were two angry bumps on either side of his forehead and blood trailed from his nostrils . . . (Pp. 403–4)

Emotions are ubiquitous in the passage, some compassionate, some angry, accompanied by bodies that deliver blows to a fellow spectator-sufferer Biafran; not just any Biafran but a (hungry) combatant on whom the breakaway region relies to make its vision of the future come true.

Berlant (2004: 7; also 2000) suggests that the West is in the grip of a "contemporary culture of true feeling that places suffering at the center of being ..." Tony Burke (2007: 1) insists that we are addicted to suffering. Elizabeth Spelman (1997) analyzes the many ways that suffering is both ubiquitous and brings out elements of indifference, contempt, and manipulation in Western cultures. Our politics can create and then pander to fear and dread, those twin-set emotions that people are so accustomed to being with as part of the fabric of everyday life. And yet, as with war scenes from Biafra, there is an irritation with observed suffering, too. There can be a desire to attack and punish, thereby breaching the borderline of spectator/ sufferer. Others can endeavor to look away and become "mere" spectators instead of compassionately obligated ones. Somewhat perversely, one can also feel personally good from acting with compassion toward others who feel bad and who will not feel good for a long time. Those lines have blurred even more in wars started by the West in Iraq and Afghanistan, where there is simultaneous engagement, withdrawal, bombing and rebuilding taking place alongside the more removed sentiments in the West about the dangers troops face (and, except for the journalistic human interest stories, not about the dangers that ordinary Iraqis can routinely experience). These messages warn against sentimentality when writing war up from people. Those people may or may not be feeling what we think they feel. The emotions of war must be got at empirically.

Mourning as politics: Judith Butler

That people can have interchangeable experiences of war – now sufferer, then spectator, or both simultaneously – and also shift within categories or types of spectatorship and suffering, underlies Judith Butler's (2004a; 2010) masterful works on the significance of war-induced mourning to a potential global politics against war and violence. This is her problematic: "what form of political reflection and deliberation ought [we] to take if we take injurability and aggression as two points of departure for political life"? (2004b: xii). It is a decidedly broad remit that crosses between war and peace and injury and politics in the now and the future. Inspired by the aggressive reaction of the United States to the attacks of September 11, 2001, Butler has been especially interested in considering the emotion of grief and how it can become the basis of less violent politics and living than the world has witnessed in the past ten or so years. She writes: "To be injured means that one has the chance to reflect upon injury, to find out the mechanisms of its distribution, to find out who else suffers from permeable borders, unexpected violence, dispossession, and fear, and in what ways" (p. xii). In other words, to suffer directly or indirectly from the loss of people close to one – or perhaps the loss of a limb – is to work the affective– emotional process into a politics. Which way that politics goes depends on whether one makes the connection between personal war emotions and a larger shared condition by asking who else suffers in wars, rather than fixating on how much we are suffering. Butler famously argues that "some lives are grievable, and others are not" (p. xiv). Clearly, people like us, those who are on "our side," are grieved openly and acceptably in war situations. So many others are suffering too, but by virtue of being on the wrong side in the war their losses are neither enumerated nor seen as shared. To Butler, it is imperative to circumvent the longstanding practice of seeing others as less human than we are and therefore as unavailable or irrelevant for political alliances.

Hers is a distinctly emotion-based, but socially and politically alert, approach for an era of new wars that target or engage large numbers of ordinary civilians as sufferers and spectators, and the state-led terror wars in Iraq and Afghanistan. Butler lays out the relationships this way. "[E]ach of us is constituted politically in part by virtue of the social vulnerability of our bodies" (p. 20). We are attached to others and when we lose them we are vulnerable to the point of being undone by the loss, gripped and altered by devastating emotions we cannot channel. Our subjectivities career about as we try to work out why the presence of particular absences is so difficult. What is it that we have lost? Why are we suddenly so vulnerable? What is our future? As upsetting as emotions accompanying loss are, the harsh and the painfully undone can transform us, Butler contends, into empathetic bodies that recognize the losses others must feel and the grief they experience. I would put it this way: the emotions of loss can send us traveling within our emotional repertoires as a prelude to traveling to others who also experience corporeal and emotional vulnerability. In a sense, Butler is also saying that the emotions around personal loss can become the basis of a compassionate way of living and politics based on collapsing the sufferer–spectator dichotomy internationally. She writes: "[m]any people think that grief is privatizing, that it returns us to a solitary situation and is, in that sense, depoliticizing. But I think it furnishes a sense of political community of a complex order, and it does this first of all by bringing to the fore the relational ties that have implications for theorizing fundamental dependency and ethical responsibility" (p. 22).

Of course, this is trickier than it might seem if one thinks of the community of loss in war as far larger than the USA or other similar cultures. Americans like Cindy Sheehan, who have lost offspring to wars in Afghanistan or Iraq, are likely to present different cultural and political manifestations of sorrow, grieving, and mourning than Afghanis and Iraqis. And there are likely to be different mechanisms for dealing with a perceived loss of oneself, of parts of oneself, as a result of a shocking war experience. Maria Eriksson Baaz and Maria Stern note, anecdotally, that women who have been raped in the Congo say they do not want the psychological counseling that various UN agencies consider as a healing mechanism. Counseling might work in Sweden or other Western countries. The women in the Congo say they would rather have money for their children's food and clothes: that is something that would make these women feel better.[7] Such is the type of cross-cultural gap I also found when I compared the work that Zimbabwean women said they wanted to do to the work options on offer there in the 1980s and 1990s from international organizations brimming with resources for women in the aftermath of the country's independence war (Sylvester, 2000). It can be problematic, therefore, that Butler draws on Western psychoanalytic traditions that are not known for cross-cultural sensitivities. The risk is that one can get caught up in implicit universalist thinking. Yet I find that Butler does see people as collaged with elements that are alike and different. She believes that we are all vulnerable "to a sudden address from elsewhere that we cannot preempt" (Butler, 2004b: 29), and yet our common vulnerabilities "cannot be thought without difference" (p. 27). Our emotions, passions, and human capacity for grief can "tear us from ourselves, bind us to others, transport us, undo us, implicate us in lives that are not our own, irreversibly, if not fatally" (p. 25). With precarious lives mutually vulnerable and mutually grievable, beyond and despite and also because of cultural difference, we "make a mistake if we take the definitions of who we are, legally, to be adequate descriptions of what we are about" (p. 25). Yet how we get transported into a political anti-war movement or politics comes across as sketchy.

A staged encounter: Lauren Berlant, Judith Butler, and Erin Manning

It might be said that Butler is in that cadre of thinkers that Berlant (2004; also 2000) decries for setting up suffering as the true center of being, and that leads Burke (2007: 1) to assert that the contemporary West is addicted to suffering. The late Gillian Rose (1996: 11) called the poststructuralist "suffering" genre "a process of endless mourning, lamenting the loss of securities which, on its own argument, were none such." As noted above, Berlant is keenly aware that compassion can awaken very different sentiments than the shared suffering Butler emphasizes:

> scenes of vulnerability produce a desire to withhold compassionate attachment, to be irritated by the scene of suffering in some way. Repeatedly, we witness someone's desire to not connect, sympathize, or recognize an obligation to the sufferer; to refuse engagement with the scene or to minimize its effects; to misread it conveniently; to snuff or drown it out with pedantically shaped phrases or carefully designed apartheids; not to rescue or help; to go on blithely without conscience; to feel bad for the sufferers, but only so that they will go away quickly. (Berlant, 2004: 9)

In other words, it is quite possible to focus more on being compassionate to oneself than to others, even others who are like oneself or face similar predicaments. What if, Berlant asks, "it turns out that compassion and coldness are not opposite at all but are two sides of a bargain that the subjects of modernity have struck with structural inequality" (p. 10)? Intriguingly, Butler (2004b) does not reference Berlant's ideas in *Precarious Life* and does not alert the reader to compassion's Janus faces. That is a pity.[8]

The critical development camp running alongside IR is very aware of the underbellies of compassion. In the development industry of powerful international organizations and nongovernmental agencies, compassion can indeed be very cold and cruel. Marianne Gronemeyer (1992: 54) notes that "the cry for help of a person in need is rarely any longer the occasion for help. Help is much more often the indispensable, compulsory consequence of a need for help that has been diagnosed from without." The development expert instructs far more often than s/he might listen. Development tempts, tantalizes, chastises and creates a desiring but always disappointing other who reaches for help and is doomed to not-receive it quite from the helper. In effect, the helped are spectators that the helpers, themselves spectators, must suffer in often brief and irritating encounters. Unlike the people Butler implicitly pictures as collaborators in living changed lives through experiences of shared war grief, many development specialists are only tangentially even in the site(s) they visit or model (Sylvester, 2004; 2011d). Many are compassionate and show an admirable sense of urgency – as they hold themselves aloof. Their compassion has bodily, emotional, social, and geospatial boundaries.

Returning to Butler, however, her important contribution, and her difference from others, is seeing beyond feelings and beyond levels of expertise – who instructs whom – to a body-based, emotion-propelling politics that could enable discursive intervention before military intervention occurs. That politics would be based on the shared sense that war is "a touch of the worst order" (Butler, 2004b: 28). Her argument intersects interestingly with Erin Manning's *Politics of Touch* (2007) discussed earlier (Chapter 3). Manning maintains that "touch is not simply the laying of hands. Touch is the act of reaching toward, of creating space-time through the worlding that occurs when bodies move" (p. xiv). Presumably, the reaching toward can be benevolent or malevolent, which means that the touch can be warming, healing, or exciting in its body–emotions effects; or the reach can move away from healing toward an injurious war touch. Butler has in mind a reaching in space and time toward others who have already experienced or are currently experiencing some aspect of the war touch and are grieving as a result. She argues that emotions of war grief shared by enemy and friend alike could be massaged into a global political community in which "suffering unexpected violence and loss and reactive aggression are not accepted as the norm of political life" (p. xiv). In other words, Butler is suggesting that suffering (war suffering at least) can be a motivating factor in reaching, interdependently, toward new norms of living that do not elevate a certain range of emotions either as normal or as regrettably collateral but also normal.

For Manning, touch can similarly operate alongside politics in inventive ways, "by drawing the other into relation, thereby qualitatively altering the limits of the emerging touched-touching bodies" (p. xiv). And might not Butler agree? For Butler, it seems, the key question is a parallel one: not exactly what the body and its emotions are in strictly ontological terms but what the body and emotions can (also) do when forced by injury into the possibility of thinking and behaving transnationally and politically. "Mourning" for Butler is what "touch" is for Manning: both a harbinger of personal transformation that can be the basis of a new politics and an "incorporeal experience of contact" (Manning, 2007: 134). Manning uses the term ontogenetic to describe the kind of body she has in mind – a body always in genesis (p. xxi) rather than an entity that is fixed and stable, and that always-in-motion body includes affective-emotional elements. "What a body can do depends on the expressions our reachings take" (p. xxiii). Indeed.

The reachings Butler (2004b) would like to see are toward grief rather than toward relief from it. She remembers US President Bush announcing on September 21, 2001: "we have finished grieving and that *now* it is time for resolute action to take the place of grief. When grieving is something to be feared, our fears can give rise to the impulse to resolve it quickly, to banish it in the name of an action invested with the power to restore the loss or return the world to a former order, or to reinvigorate a fantasy that the world formerly was orderly" (pp. 29–30). At the end of grief comes the clamor for war, unleashing another round of grieving as some body-coffins return home buried under the flag of national sacrifice and other bodies go nameless and ungrieved in some threatening place "over there." Butler thinks there might be "something to be gained in the political domain by maintaining grief as part of the framework within which we think our international ties" (p. 30). Meanwhile, Manning (2007: 136) reminds us that "bodies reaching toward are abstract bodies," in the sense that they are always becoming and never complete that process. The question both Butler and Manning cannot answer, therefore, is how the grieving-reaching becomes a transnational politics with the power to effect pauses in other feelings, and to prevent reachings excited by fantasized experiences of wars of retribution.

But what of war's enthusiasms?

Throughout this discussion, the emphasis has been on bodies and emotions accompanying hurt and injury (Butler) and movements and experiences that can assist in unraveling security arrangements that link tightly to violence and war (Manning). What of bodies and emotions that express or reach toward enthusiasm or satisfaction in war? Are they gripped by false consciousness, inauthentic emotions, sick affects? Are those such rare emotional experiences of war that they need not be discussed? Asked differently, if "[e]xperience is animated by the senses, sensing," as Manning (2007: 141) suggests, then what of those senses that inspire the potentialities and actualities of war and not the aching sense of loss that animates Butler's discussion? What if one is quite willing to experience the chaotic thrill of the moment rather than the long-term calm of peace? In order to talk about war as experience, it is necessary to come to terms with experiences that do not culminate, have yet to culminate, in broad reaches away from war.

There are relatively few studies of happiness, joy, or satisfaction and war, but there are studies that speak of a phenomenon called "combat flow," which produces such absorption in a situation that a soldier can feel a sense of satisfaction and even well-being in combat that is strong enough to be addictive. War historian Yuval Noah Harari (2008: 253) describes combat flow as the experience of full concentration in the present moment and even emotional exhilaration, because "goals are immediate and clear; action and awareness merge; reflective and self-consciousness is lost." Others add that flow entails losing awareness of oneself as a social actor; having no irrelevant thoughts, worries or distractions; no sense of time; and a feeling that one is operating at full capacity (e.g., Nakamura and Csikszentmihalyi, 2002). Popularly known as being in the "zone," a feeling familiar to academics and other writers when ideas and words pour onto the page, flow is associated with overall well-being. One Vietnam veteran describes how the end of the war disrupted the experience of flow in ways that other activities could not recover: "Within a year I began growing nostalgic for the war ... I could protest as loudly as the most convinced activist, but I could not deny the grip the war had on me, nor the fact that it had been an experience as fascinating as it was repulsive, as exhilarating as it was sad" (Caputo, 1977: xiv; also Lloyd, 1999; Tolstoy, 1957).

Methode's mode of dying in *A Sunday at the Pool in Kigali*, mentioned in Chapter 3, could be interpreted in terms of flow as the last hurrah of a man suffering what Giorgio Agamben (1998) would term bare life. In his context, Methode *is* doubly bare – as an AIDS victim and as a Tutsi in an anti-Tutsi genocide. But even in his bare life circumstance, damned either way it seems, Methode resists the predefinitions of approaching death – misery, suffering, and-then-you-die. As Manning might foresee, he reciprocally reaches towards his trusted friends and in-gathers them and the world, even as he worlds them to a rewritten performance of dying. Another biopolitics, one Methode authors, substitutes for the one he is meant to enact. He also refuses to let the Rwandan state kill him with its genocidal governance strategy, and, instead, embraces, in effect, the pernicious international epidemiology of AIDS. This is not anyone's idea of an ideal trade-off, but it is thought flowing intensely, in motion. His very last moments of sex, morphine, and whiskey put him in a joyous zone of pleasure where one would not expect it. Methode's insistence on a different method of apprehending life and impending death shows a grey area of living that celebrates selected motions of life within bare life, reaches towards these, embraces them, and does so in the company of others who help him move the relations of life into a death-flow saved from ugliness. As another character in the book says to a Westerner who is trying to help: "Leave us to die peacefully alive" (Courtemanche, 2003: 82). That is Methode's method.

There is also the kind of joy in war that can grip people who start a glorious war against oppression. Adichie's *Half of a Yellow Sun* (2007) is full of death-rattles and yet it is also a tale of the enthusiasms and unflinching convictions that lead a group of people to believe their cause is so just that it will prevail despite blockades, air attacks, starvation tactics, mounting fear, and increasing evidence of defeat engulfing them. The collective delusion that one region of Nigeria could, with some military equipment plus home-made rockets and grenades, defeat a nation-state backed and armed by Britain takes a long time in the novel to dissipate – way past the time the blockade reduces incoming food and petrochemical supplies to a minimum, past the time hospitals run out of drugs, and even after so many people's hair turns red from malnutrition, and their children's hair too, and falls out by the fistfuls. It seems so easy in the beginning. Professor Ekwenugo entertains his friends with news of the weapons his Science Group is making for the war on the "vandals," the Nigerians of the north:

"We launched it this afternoon, this very afternoon," he said . . . "Our own home-made rocket. My people, we are on our way."

"We are a country of geniuses!" Special Julius said to nobody in particular. Biafra is the land of genius!"

Ugwu [a thirteen-year-old houseboy] sang along and wished, again, that he could join the Civil Defence League or the militia, who went combing for Nigerians hiding in the bush. The war reports had become the highlights of his day . . . the guests sang and shouted drunkenly about the might of Biafra, the stupidity of the Nigerians, the foolishness of those newscasters on BBC radio . . .

"They are surprised because the arms Harold Wilson gave those cattle rearers have not killed us off as quickly as they had hoped!" (Pp. 198–99)

In Adichie's close-up of Biafra, we can see that the grand delusion is maintained mostly by those of privilege and education, who are not actually fighting what starts out as the promising war and ends in tears. And that delusion does not last. War becomes a gaping injury to spirit, body, convictions, emotions, and honor: you return to the relief center the next time wearing a rosary around your neck "because Mrs. Muokelu said the Caritas people were more generous to Catholics" (p. 283). Humiliation, numbness, and grief – these emotions highlighted by Adichie tend to stay within oneself rather than extending to other injured Biafrans, to say nothing of the injured Nigerians – the enemy Biafrans are interacting with intensely.

Mixed emotions and war

As at the conclusion of the previous chapter on physical bodies, the mind does not come to rest on any definitive sense of emotion here. Research in the area of emotions is newly increasing by the minute. Lawrence Grossberg (2010: 317), a leading figure in the development of cultural studies and of affect studies, has this to say about the new interest: "There is an important question of why 'affect' has been so consistently ignored, along with other concepts like emotion and the body, in the dominant traditions of Atlantic modern thought. I think that part of the answer is no doubt, as feminists have argued, the association of women as somehow inferior with the assumption that the sexual difference manifested itself through a series of binary differences: rationality versus emotion, mind versus body, and so forth." He goes on to say that more is involved than gender power relations, though, such as recent confusions about what emotions are relative to affect; also, research in this area has interrupted dominant traditions. He himself is of the school of thought that says "emotion is the articulation of affect and ideology. Emotion is the ideological attempt to make sense of some affective productions" (p. 316).

Throughout this chapter, feminist analysis has largely seen through the gendering of emotion in IR and has moved productively into critical positions, some borne of research in war and conflict zones, as in the case of several participants in the Forum on Emotion and the Feminist IR Researcher. Whether emotion is an ideological attempt to make sense of affect is an open question. Connolly calls attention to a bodily relay of affect and a working-out of what that means socially; rather than arguing that emotion involves ideology, one could more simply say that it involves the body processing diverse sources and senses of knowledge. That would mean that arguing against the body as a site of emotion would be off the mark. Given the new scientific research on body–mind relations, theorizing away the body as a key site of emotion, as some of the analyses mentioned here advocate, is where ideology would really enter the picture. In effect, that type of theorizing is the ideological move. At the same time, arguing against external triggers of body-producing feeling is equally one-sided. Butler's (2010: 50) work on war and emotion inclines in its recent forms toward the emotion-as-ideology argument. Yet her approach to cultural criticism per se is an arrow straight into the heart of war as an institution. She puts war and criticism together in a new frame of analysis that centers "the emergence and vanishing of the human at the limits of what we can know, what we can hear, what we can see, what we can sense" (Butler, 2004b: 151). As members of social communities, we become inaugurated into commonly accepted meaning and sense systems. Just as we are initiated into "proper" gender identities, and just as Bordo lambasts the social acceptance of feelings that one's body is inadequate, so Butler (2010: 51–52) argues that "war sustains its practices by acting on the senses, crafting them to apprehend the world selectively, deadening affect in response to certain images and sounds, and enlivening affective responses to others." She does not say this quite, but it is clear that to do this, war must indeed be a social institution that influences people's views of it, their acceptance of it, and their sense of whose life can emerge from it and who is meant to vanish into its fog.

Enloe might argue here that war is the ideology and gendered emotion is one of its tools; but that is an unstable and risky terrain on which to place all bets. Emotions do not always support war. In not quite acknowledging counter-tendencies of body–mind relays, Butler's otherwise excellent analysis becomes too deterministic, even reductionist – too social, perhaps, to the point of causing the neuros, relays, affects, emotions and all to fall down before the scripts of war. Do they fall down before war in the West and elsewhere? Do war feelings hold steady or do the people of Biafra, who start out lauding their hopeless war, remain pro-war while later wearing their rosaries around their necks to the relief center? Surely spectators, enthusiasts, and sufferers alternate places or move into new spaces of interpretation as their senses of war – their emotional and bodily reactions to it – change. Yes, "one's body is never fully one's own, bounded and self-referential . . ." (Butler, 2010: 54). But neither is it fully at the mercy of social tendencies toward militarism or social institutions like war. Rwanda had the most God-awful genocide in 1994, and earlier episodes of similar wars. Where is it and Rwandans now? Pro-war? Anti-war? Full of mixed emotions?

These are empirical questions that study-up approaches to war can explore: they relate to how experiencing war affects people and is affected by them. It strikes me that Butler, whose writings I greatly admire, also needs to study up rather than study war down mainly from a US-centric understanding of its war emotions and enthusiasms. People in war zones are likely to have body-based emotions that do not validate the sophisticated arguments Western theorists construct about the social conditioning of the emotional. When one encounters war directly rather than indirectly, what exactly is the social script programmed into the body? There is none. A woman living in Berlin through the city's siege ending World War II writes in 1945:

April 21
Bombs made the walls shake. My fingers are still trembling as I hold my pen. I'm covered in sweat as if from heavy labor. Before my building was hit I used to go down to the shelter and eat thick slices of bread and butter. But since the night I helped dig out people who'd been buried in the rubble, I've been preoccupied, forced to cope with my fear of death. The symptoms are always the same. First sweat beads up around my hairline, then I feel something boring into my spine, my throat gets scratchy, my mouth goes dry, my heart starts to skip. I've fixed my eyes on the chair leg opposite and am memorizing every turned bulge and curve. It would be nice to be able to pray. (Anonymous, 2000: 10–11)

June 15
Sometimes I wonder why I'm not suffering more because of the rift with Gerd [her erstwhile fiancé newly arrived from the East], who used to mean everything to me. Maybe hunger always dulls emotions. I have so much to do. I have to find a flint lighter for the stove: the matches are all gone. I have to mop up the rain puddles in the apartment; the roof is leaking again; they merely patched it up with a few old boards. I have to run around and look for some greens along the street curbs and stand in line for groats. I don't have time for feeding my soul. (P. 261)

Different proximities to war matter. Rather than jumping into the social-dominant end of the pool, I prefer to think of bodies with emotions swimming with and against conventions depending on their experiences of all sorts, including their experiences with the transhistorical and transnational, ever-changing but continuously injurious, social institution of war.

5

CONCLUDING, COLLAGING, AND LOOKING AHEAD

This book began with war in our time as a profusion of collectively violent activities that involve a great number of people and also a wide range of authorities and types of experience. Some IR analysts claim that war is declining and the idea of peace is gaining in the world (e.g., Goldstein, 2011). I do hope that is the case, but I am increasingly convinced that war might be waning in statistical terms but not in human terms. The presentation of aggregate statistics on war types and frequencies, coupled with war end-dates that correspond to peace treaties, tells very little about the experiences of war, the people who are affected for a long time by wars in direct and indirect ways, and the impressive tenacities of an ancient social institution that shape-shifts through time and is with us yet. The types of data that support the thesis of waning war rely on standard IR methods put in the service of standard scientific generalizations. That approach is not adequate on its own for investigating war as experience and the myriad people who suffer or observe it – not in the first instance, at least. Experiential "data" is more likely to come from slogs through the fields or texts of war or post-war, gathering stories and crafting narratives, considering grey zones of experience of all sorts and time-frames that may or may not accord with official views on when wars end; having done that, a researcher might then wish to look for patterns of similarity and difference across war experiences and/or zones. A quantitatively oriented researcher might want to flesh out the usual statistical curves and get more people associated with the data that are reflected there. Approaches for studying up and for studying down should meet, for the benefit of war studies and all people in the war matrix.

The academic field of IR knows war, but it is not very good at knowing people or at putting social relations into its war frameworks. Indeed, much of IR can seem to have that aversion to people that the feminist anthropologist Lauren Berlant (2004) decries about the social sciences. We train our students to avert their eyes when it comes to people, bodies, emotions, and the enormous palette of war colors and forms that could, in fact, be at the crux of war rather than at its periphery. To put this matter another way, much of IR has a spectator role *vis-à-vis* war and rarely sees itself as one of war's participants. In his inaugural speech as President of the International Studies Association (ISA), Steve Smith (2004) suggested that the discipline of IR was both in thrall to and partly responsible for global violence of all kinds. In its pristine separateness from the fray, via the insulations required of value-free social science approaches, and in its periodic involvements in war politics and policy via scholarly consultancies, IR could be considered a war collaborator. That is to say, the field is complicit with the social institution of war. In Smith's own words:

> all of us in the discipline need to reflect on the possibility that both the ways in which we have constructed theories about world politics, and the content of those theories, have supported specific social forces and have essentially, if quietly, unquestioningly, and innocently, taken sides on major ethical and political questions. In that light I need to ask about the extent to which International Relations has been one voice singing into existence the world that made September 11 possible. Please note, I mean all of us engaged in the discipline, not just "them," whoever they are, not just "the mainstream," whatever that is, and not just "the U.S. discipline," however that is defined. Rather, I will take this opportunity to ask each and every one of us about our role in September 11, and thereby to reflect on the link between our work, either in writing or teaching, and international events. (P. 500)

This wording does not demonize any particular branch or camp of IR, nor does it offer kind words for some and not others. That means feminist IR is not off the hook in Smith's argument, even though it has been known to portray itself as a morally superior but often misunderstood waif of IR. Critical theories and postcolonial theories that can also seem above "backward" IR are not given a break in this statement either. We are all in this together, Smith is saying, and "this" IR is not a community of glad tidings, peace, or cooperation. It is war's collaborator.[1]

Throughout this book I have endeavored to argue that IR must study war with greater direct interest than it has in recent years, and with more emphasis on the experiences of war, rather than pursue any single-track focus on abstract actors and mechanics. War is defined here in terms of its massive capacity to injure people and communities. People are war's perpetrators, its targets, and its very content. The sophisticated weapons of today kill and maim in ever ghastlier ways and also keep company with machetes and crude homemade IEDs (improvised explosive devices). Why are we in IR averting our eyes from the social intensities of war? Why are war's abstractions so fascinating to IR in a time when it is the people of the world who are taking the bullets and the body abuses that come both from being in war zones and from massing peacefully to demand freedom? Why is feminist IR so late in coming to the study of war at all, let alone as a peopled institution, set of practices, and arena of opportunity? Social anthropologists, novelists and war journalists, among others, have been way ahead of all of "us." Feminist aloofness from war studies until recently, owing to the conviction that war is someone else's misguided activity, and its ethical belligerence toward war in the name of peace, can be read as avoidance and denial of ugly truths of violence around the world. And it is an avoidance that can be carried off with a righteousness that feminists find so offensive when it comes from those Smith calls "them." Smith is right: we can all do better than this. If there is one thing I hope this book demonstrates, it is that people show up with war agency and war wounds. They are the wars, and it would behoove us to cast our gaze in their direction rather than turn away.

Overlapping possibilities and linkages

One of the joys of researching a book on war, feminism and IR is that it throws up so many overlapping interests, subterranean and explicit, between camps that rarely acknowledge one another. It is fascinating to discover that realist theorists like John Mearsheimer and Stephen Walt can rely so effectively and openly on the presence of people and their injuries when seeking to influence Washington's policy toward Israel. Read their other more abstract realist works and you might wonder whether the world is populated at all; perhaps it is empty save for Western elites. Read James Der Derian's (2009) riveting interview with Admiral Cebrowski at the US Naval War College and find that in Cebrowski's view, "network-centric warfare is behavioral-based. Many folks miss that. They go to the 'network,' which is the adjective; warfare is the noun. Warfare is based on human behavior" (p. 135). The Admiral's idea of human behavior and "behavioral-based" might be different from yours or mine, but it is there, nonetheless: the human. Of course, Der Derian's brilliant book contains nary a woman and offers zero feminist thinking; but like some of Mearsheimer's and Walt's work, it has other strengths that feminist and other frames of war studies can build into new and more encompassing understandings of war.

Smith argues critically that IR is usually on the search for similarities, general patterns, and fixed propositions. Yet feminists can be oriented that way too, when they talk about masculinity, patriarchy, and militarism as *the* culprits behind war; suddenly the world is a place of great similarity, despite even ardent metanarrativists recognizing that feminist analysis today is about accepting and exploring difference, and about finding intersections where valorized particularities once reigned. Feminist IR analysts cannot balk at the different ways IR clandestinely brings people into studies that do not purport to "do" people at all and also claim feminist sensitivity to difference. The dilemma I pointed out earlier is this: whose and which differences count as real differences? Is Hirsi Ali a legitimate difference within Islam that feminist analysts should acknowledge? Is Mearsheimer a legitimate difference from feminist IR within IR? The terrain of difference is rocky, to be sure, but feminist students of war could look for odd points of entry that can be pried open and spilled into a large war problematic, working with rather than against or in isolation from camps of great difference to it. Ditto in reverse: IR's camps should be much more open to opposing camps than they are.

It is encouraging, of course, that war does turn up in feminist IR now where it was once not encouraged to be. The recent research reviewed here is mind-boggling in its contributions, not the least being a gutsy determination to experience war directly or indirectly as part of studying it. We have seen scholars like Miranda Alison, Maria Eriksson Baaz, Megan MacKenzie, Swati Parashar, and Maria Stern – all identified with feminist and IR research – presenting the field with enormously important insights on war from women combatants in Sri Lanka, Northern Ireland, and Sierra Leone, by male combatants in the Democratic Republic of Congo, by militant women in Kashmir and by people's violent anti-state organizations in India. We have also seen feminists exploring the narratives of women's experiences with the securities and insecurities of war in the Indian partition, the Balkan wars, in Mozambique, Palestine and many other places. The willingness of this generation of feminist IR scholars to take to the fields of war and post-war is something methodologically new to an abstraction- and distance-enamored IR. It shows an open embrace of methodologies more common to social anthropology than to IR. In fact, dare it be said that some feminist IR war studies merge with epistemological and ontological components of social anthropology. Perhaps that is exactly the kind of move required if one agrees with Smith's sense of what IR is missing: "The puzzles and conflicts of international relations cannot be 'solved' in the manner of a logical puzzle; instead they have to be unwrapped and understood from the viewpoints of the actors involved" (p. 511). That is what feminist IR war studies often endeavors to do, whether through ethnographies, textual analyses, or other methodologies.

Where to go next

The main argument of this book is that "the actors involved" with/in war are people, often ordinary people that a field has excluded in the past under the assumption that such people are not stakeholders in international relations. I want to take these final pages of the book to push that argument further and take it in several directions, even though it means introducing a few new ideas and literatures at the eleventh hour.

I propose first that the collective violence we associate with old and new wars is, definitionally, experiential in nature. That proposition follows from Elaine Scarry's (1985) persuasive argument that the content of war is human injury, which is an experiential realm of politics. War is about human injury and human injury is experiential, which suggests that war is and should be studied as an experiential realm of violent politics. Of course, it is important to avoid building experience into a new metanarrative that can explain or capture *the* essence of war: there is no one essence. War has many components that can be pulled out for theoretical analysis – gender, technology, ideology, militaries, weapons, histories, and so on. And as a second and related matter, I want to argue that one important component of war is the much-maligned state. The perspective of war as experience reduces the centrality of the state to war, and thereby also reduces its singular importance in international relations and IR; but the state cannot be consigned to insignificance. Feminist theorist Wendy Brown (1995) depicts the state as a site of injury and injuring too, further suggesting that the social world of war revealed by studying up from people must be linked at some point with the views down from the state, its bureaucracies, and its murky political economies of war. New war thinking has usefully exposed the privatization and outsourcing of state military functions and personnel, a topic that several IR analysts fruitfully pursue in their research (Abramson and Williams, 2010; Leander, 2012; Stanger, 2009). But Ronnie Lipschutz points out that many neoliberal states tout the benefits of reducing the state through outsourcing, when, in fact, such states actually grow in size through intricate political economies of power-sharing.[2] Thus, Obama brings the troops home from Iraq in 2011 and carefully avoids telling the public that scores of private military operations contracted by the US will continue on there.

As well, we all know of the dramatic rise in political economies of surveillance and policing in states that are either preoccupied with security in the post-9/11 era or, as some of the affect studies scholars point out, find it useful to stir up public fears in the pursuit of myriad other, sometimes anti-democratic, agendas. Indeed, descents into the ordinary of violence and war can bring one into close encounters with state technologies of development (Duffield, 2007), with state peacekeeping "warriors" operating in conflict zones (Kronsell, 2012), and with gendered states everywhere. The prerogatives of liberal states are vast, but states can be powerful even in cases where they fail to protect their territory and population from marauding gangs, terrorist sects, or everyday conditions of impoverishment. Strong or weak, vulnerable, failed or transitioning, states are still players in state–interstate, state–corporate, state–NGO–IGO, and state–people relations. Yet overly tight and overly rehearsed principles of state-centric thinking are inappropriate today, belied every day by the actions of many other authority centers of war and by experiences in not-at-all-abstract conditions of war that might not feature state militaries so much as warlords and their forces.

As a third point, the pervasive assumption of rationality in international relations is also a stumbling-block, belied by the evidence of neuroscience that rationality does not exist as it is used in IR and never exists independently of emotions. Rationality is a body–mind capability, but it is not the only or necessarily the key one. As Smith is at pains to point out, rationality is embedded in many branches of IR; it is a realist point of faith, but that faith is shared with liberal models of aggregative and deliberative democracy that deny, as Chantal Mouffe (2002: 8) puts it, "the central role of 'passions' in the creation of collective political identities." Adjustments to state-centric theory and the rationality premise could reposition some camps of IR just enough to enable productive conversations with feminist scholars and others about locations and experiences of violent politics at the borders of the domestic and the international. That is where humanitarianism has been "rationally" attached to war, creating the perverse kill-to-be-kind illogic of humanitarian intervention (Orford, 2003; Sylvester, 2005).

A cutting-edge search for linkages across levels of analysis is a prospect that even Smith does not entertain as a way for IR to sing the world into existence. But it is a fourth step to take. Perhaps critical theorists focus so much on the fallacies and presumptions of realist and liberal IR theories that we do not register the interesting anomalies that can exist within traditions of thought with which we have always disagreed. We can fail to notice the people brought to the surface of some IR arguments in order to enliven and strengthen abstract points; the people that Mearsheimer and Walt (2007) momentarily allow to peep out and get recognition are integrated into realism and therefore into IR – just not on their own terms. Those people have to be unwrapped and understood from their perspectives as well as via interpretive mediations of their narratives. By expanding Shapiro's ideas, it could be argued that "personal" households in bloodied areas of the Middle East are sites of everyday body-space movements that comprise micro–macro confrontations and collaborations. Point five, therefore, is that the relays of which William Connolly (2002) speaks, and the discontinuities of role and identity that war experiences are likely to produce, as claims Shane Brighton (2011), are where linkage-oriented research efforts should also focus. Such would mark a new feminist-, critical-, postcolonial-inspired IR war studies that unwraps people and connects them to one another, to various private interests around them, and to their warring (or not warring) states of the present and past. Double linkages would be ideal, stretching horizontally across the camps of IR and vertically from people to states and back down again (without averting the eyes along the way).

A further sixth contribution can be made by tracing the multiple paths and relays of authority that produce certain types of participant and spectator war experiences. As Brigitte Holzner (2011) recognizes in the case of the Liberian war, women of peace and women of war can work simultaneously with and against the war authorizing centers near them. Black Diamond and her all-women squad of fighters challenged the masculine authority of guerrilla forces that had abducted them and then turned them into bush wives. Her fighters became a force those men feared. Meanwhile, the women of peace, who were recognized for their achievements by the Nobel Peace Prize committee in 2011, worked to end the war Black Diamond and others were fighting, by becoming a force of such authority for peace that the head of state, Charles Taylor, himself a major authority in the war, finally had to acknowledge and, in a sense, obey it. The women killers and peace women had differing experiences of war, class, age and social respect, but both took some command of the local politics of warring. We do not often imagine such scenarios, where the international community might stew over steps to take in a war situation while effective politics, deadly and peaceful, are operated by women, mostly without outside help. As a seventh point, such are the kinds of linkages that feminist IR war studies is well placed to bring to the fore: the grids of ordinary power located where others do not look. But again, there is more to do.

Enter grey zones

In the concluding chapter to *Experiencing War* (Sylvester, 2011a), I mention grey zones of war, where what happens is difficult to tag ethically, assign a category such as sufferer or spectator, or even fully absorb as a researcher or war participant. Grey zones are full of ambiguity and ambivalence for researchers, as the Forum on Emotion and the Feminist IR Researcher (Sylvester, 2011b), discussed in Chapter Four, reveals. How much more grey are war zones, where, as Mats Utas (2005) tells us, people can embrace a tactic or masquerading form of agency that adapts to situational ethics, a move that just might ensure a person's survival.[3] The researcher hard on the path of experiential information can be discouraged by such tactics and their ambiguities. Someone is lying, telling untruths, making up stories. She can also be sent on her way by faces and voices that express no affect and are willing to tell no stories. Bodies, always contested entities, can become bewildering in their multiplicities and overlapping identities during war. Anthropologists and feminist students of IR approach embodied persons to find out what happened to them – how did her or his body operate and process information, physically and emotionally, in a war zone or spectator place? The words spun around those bodies, and the emotions a researcher hopes they will share, have been crucial in answering research questions about war and experience. But what if words fail respondents? What can seem dramatic and explosive from a distance – War! – turns into an eerie grey zone where understandings of oneself and others change rapidly and many times over.

Scarry (1985) argues that pain shatters speech and language; it is therefore unsharable. She claims that "physical pain does not simply resist language but actively destroys it, bringing about an immediate reversion to a state anterior to language, to the sounds and cries a human being makes before language is learned" (p. 4). Pain, in other words, is a private bodily feeling that cannot be known entirely by others through speech. Aspects of pain can be communicated, she thinks, if the sufferer projects that pain to the world through imaginative projects such as art and literature. Veena Das (1995) does not agree. Her view is that pain is eminently communicable. Feminist Holocaust scholar, Robin May Schott (2011: 16), summarizes their contrasting views: "For Scarry, the trajectory of pain is from radically subjective pain, which in its intensity destroys language, to its reconfiguration in language by the sufferer, witness, or artist, to its subsequent communicability to others. For Das, the trajectory is from private pain to a public articulation of pain, enabled by institutional structures such as therapeutic spaces and tribunals, to pain becoming experienced in other bodies as well." Unless the researcher imposes homogeneity on people, it is more than possible that both conditions can prevail side by side, with some people unable to speak of pain and others able and willing to articulate the pain of war – or its "flows" – publicly. This is another key point to bear in mind for studies of war that aspire to reach toward the bodies of people experiencing war: it is important that the researcher hears or touches war experiences and does not just reiterate one's own theory as a local story (Spivak, 1988).[4]

Although fieldwork interviews have been embraced by many feminist IR students of war and are, in my view, one of the most promising ways to experience oneself and others experiencing that social institution, there are other ways to study war experience. Dubravka Zarkov's (2007) research on bodies of the Balkan wars pieces together narratives of gender and rape from media sources. Jill Gibbon (2011) draws people she observes at arms trade fairs, providing an unusual glimpse into experiences of war that involve buying, selling, sexualizing, and glamorizing weapons. P.W. Singer's (2006) studies of child soldiers rely on primary and secondary sources from aid and rehabilitation workers as well as sources from academic works in fields as seemingly disparate as pediatric medicine and military history; his probing work on robotics and war (Singer, 2009: 12) draws on everything from "musty old history books that hadn't left the library in years," to professional journals going back 20 years, to interviews with robot scientists and weapons developers. The war anthropologist Heonik Kwon (2010) has also found a combination of methods useful for studying the long and unfinished war in Vietnam. When reading his work, I paused over a letter he quoted from a prominent war veteran to the revolutionary Communist Party after the shooting ended: "It is according to the principle of our revolution to share wealth and happiness with generations of war dead who knew nothing but poverty and suffering. The nation's prosperity should benefit all generations of Vietnam, not merely those who are alive" (p. 91). Ancestors are undead, but not in the ways the zombie, vampire, and werewolf freaks of the West fetishize. Like Das (1995), Kwon's methods help him recognize that war pain can create community. It is that war-created community in the making everywhere that Butler discovers philosophically and is so avid to link with and share emotionally and politically.

Fact–fiction insights into war

There is also the possibility that a researcher can work with war testimonials and novels to unwrap people, war and experience. I have done so throughout this book and others, including Kwon, Schott and Singer, do so too. Kwon (2010: 97) notes, for example, that there is "a growing intellectual movement in Vietnam, among writers in particular, that is attempting to reconfigure the history of the American war, using mainly fictional and poetic forms of communication, from a coherent, unified historical narrative of self-determination to a relatively less coherent, divergent experience of a "domestic conflict." He goes on to explain that "if the war was a communal conflict as well as an international one, that means that the community should play as active a part as state actors and international organizations in bringing these conflicts finally to end." Singer devotes nearly 20 pages to the science fiction that inspires robot designers and their military advocates, paralleling James Der Derian's (2009: xxxviii) statement quoted earlier, that his own scholarly book on virtuous/virtual war is a detective story; Singer recounts US Admiral Michael Mullen proudly describing "how the navy's 'Professional Reading' program, which he helped develop to guide his sailors, includes the science fiction novels *Starship Troopers* and *Ender's Game*" (p. 150). Robin May Schott finds the anonymous diary of *A Woman in Berlin,* which I too have referred to and quoted, essential reading on emotions and war.

The words on the pages of such writings present bodies, emotions, and ambivalences difficult for researchers and other war spectators to conjure or register on their own. Those words depict the shallows and depths of violence and injury, the unexpected mundanities and inconveniences, the bits of gallows humor that can hit one in war, the moments of joy, and the contexts in which particular wars unfold. Aside from identifying war with physical and emotional injury to humans, which does not seem disputed in such sources, one enormous benefit of reading fiction and testimonials is that these discourage tendencies to generalize war: wars are alike and they are different in important details. Azar Nafisi (2004: 157–158), for example, explains in her memoir of the Iranian Revolution, *Reading Lolita in Tehran,* how the war between Iran and Iraq was initially apprehended and experienced by some Tehran residents. The fact of a war launched on top of a local revolution comes as a surprise and an omen, like turning on the TV and watching a commercial passenger jet disappear into one of New York's twin towers. Here is her spectator point of view:

> The war came one morning, suddenly and unexpectedly. It was announced on September 23, 1980, the day before the opening of schools and universities . . . It all started very simply. The newscaster announced it matter-of-factly, the way people announce a birth or a death, and we accepted it as an irrevocable fact that would permeate all other considerations and gradually insinuate itself into the four corners of our lives. How many events go into that unexpected and decisive moment when you wake up one morning and discover that your life has forever been changed by forces beyond your control . . . Down the slope of the street, on the wall to my right, in big black letters, was a quotation from Ayatollah Khomeini: THIS WAR IS A GREAT BLESSING FOR US! I registered the slogan in anger. A great blessing for whom?

For the Islamic Revolution, apparently: "Our ambivalent attitude towards the war mainly stemmed from our ambivalence towards the regime . . . [which] recommended that women dress properly when sleeping, so that if their houses were hit, they would not be 'indecently exposed to strangers' eyes'" (p. 160). And worse: "Immediately after the bombs fell and before the ambulances came, six or seven motorcycles arrived from out of nowhere and started circling the area. The riders all wore black, with red headbands across their foreheads. They started shouting slogans: *Death to America! Death to Saddam! Long live Kohmeini! . . .* Some tried to go forward to help the wounded, but the thugs wouldn't let anyone go near the place. They kept shouting, 'War! War! Until victory!'" (p. 211, emphasis in original).

Contrast that war spectator account with Nora Okja Keller's novel called *Fox Girl* (2002), the story of two girls in Korea, one born of Korean parents and the second a product of a Korean mother and an American GI. For both, the injuries of war go on and on, without any beginning they can remember in their own lives, and seemingly without end in the future. It is Hyun Jin, the puzzled and ambivalent girl of Korean parents, who narrates the struggles of coming of age in grey war–no war circumstances. In the passage below, she learns of an intricate past masquerade of war that still plays out in her family, with her at the uncomfortable center. Her "mother" is speaking:

> "After the war, I returned to Paekdu. I had a mother, a father. A little sister. Two brothers. I thought I might be able to find them back home. But there was no one. No village. No home. Our house and fields were burned to the ground … I sat in the ashes of my family's home for two, three days, maybe longer. I didn't know what else to do, where else to go. I think I would have died there, if your father hadn't shown up. Together, we decided to make our way south; somehow it was easier for us to keep moving, to keep busy, to keep not remembering … coming south, your father and I helped each other when and if we could. I guess it was habit that made him turn to me when he had trouble making a baby with his wife … I carried his seed. He gave me some money … I had to fight to get her to take you in … Back and forth we pushed you until she shoved me backward and shut the door in my face. There I was holding this newborn" – Duk Hee cradled the shirt like swaddling. "What was I gonna do? I left you at their door."
> (Pp. 117–119)

An emotionally injured Hyun Jin confronts the woman who is supposedly her mother, sparking a furious fight in the household. Jin gets thrown out and into the world of American GIs, who rehearse a sense of entitlement to local girls that the Japanese military initiated during World War II, and the status-of-forces agreement marking the ceasefire of the Korean War – but not the end of war – continued (Moon, 1997):

> From far away, the real me watched them open the shell of my body, ramming and ripping into every opening they could. I watched them spread the legs open, splitting the inner lips wide enough to fit two of the men at the same time. I watched them bite at the breast and *poji* till they drew blood, and saw them shoot themselves into and over the belly, take breaks, then come at it again. I watched, and, other than a vague sense of pity, didn't feel a thing.
> (P. 154)

Which piece of literary writing is more accurate, more truthful – Nafisi's or Keller's? Martha Nussbaum (1990) recalls that the early Greeks did not distinguish between dramatic poetry and what we think of today as philosophical inquiry. Both, she says, "were typically framed by, seen as pursuing, a single and general question: namely, how humans should live … [Moreover, f]orms of writing were not seen as vessels into which different contents could be indifferently poured; form was itself a statement, a content" (p. 15). Thus, "[t]o attend a tragic drama was not to go to a distraction or a fantasy, in the course of which one suspended one's anxious practical questions" (p. 15). Rather it was an occasion, in broad daylight instead of darkened theater, to deliberate with one's community about the key ethical issues of the day. The tragedians were teachers as much as story-tellers. That refusal or inability to draw a line between factual and fictional forms has, of course, been lost for today's social sciences and much of Anglo-American philosophy. Gathering facts through materials and sources deemed factual is a sign of methodological soundness and ability to add to knowledge produced by other fact-conscious approaches. Fiction, a pursuit of "humanities" or "arts," presents forms and rules of interpretation that seem to stray far from the path of truth-seeking and explanation. It is all make-believe. But is this so?

Can one read Nabokov's *Lolita* in Tehran, in the middle of an Islamic revolution, and garner pointers on living life there, at that time, and in a distant culture from the one the author portrays? Or is the pursuit of the book under conditions of serious social upheaval trivial, whimsical, and even incorrect from a postcolonial point of view? Is the reading of it worth the effort, the political risks entailed in meeting seven or so women students secretly, clandestinely in Nafisi's apartment, as though plotting the overthrow of the government and ruining its "blessed" war? Here is what Nafisi says:

> Take Lolita. This was the story of a twelve-year-old girl who had nowhere to go. Humbert had tried to turn her into his fantasy, into his dead love, and he had destroyed her. The desperate truth of Lolita's story is not the rape of a twelve-year-old by a dirty old man but the confiscation of one individual's life by another. We don't know what Lolita would have become if Humbert had not engulfed her. Yet the novel, the finished work, is hopeful, beautiful even, a defense not just of beauty but of life, ordinary everyday life, all the normal pleasures that Lolita . . . was deprived of . . . I added that in fact Nabakov had taken revenge against our own solipsizers; he had taken revenge on the Ayatollah Khomeini . . . [and those who] had tried to shape others according to their own dreams and desires . . . Nabakov, through his portrayal of Humbert, had exposed all solipsists . . . (P. 33)

Take also Rebecca West, writing at length and controversially about her 1937 journey through Yugoslavia and her impressions of the regional people. It is a naïve, romantic, personal and contradiction-riddled analysis of a complicated area of the world. Nonetheless, it is a key text that the Clinton administration will pick up 50 years later as a valid analysis of ancient ethnic hatreds; it will influence the US government to abandon American intervention in the Balkan wars in 1993. Lene Hansen (2011) has newly rendered this enormous tome relevant to IR, rereading it with an eye to how West conceptualizes the international and the political and how her difficult-to-take-onboard observations illustrate aspects of a feminist consciousness working to understand imponderables.[5] She deems it a classic in its hybrid genre of travel writing, political treatise, and feminist thinking. West pens quite long passages on the imperial histories of the region and on what she sees as Britain's irresponsible refusal to intervene to save the area from German encroachment in later years. She can idolize Serb men and present patronizing views of women; but she also shows a strong sense of the ambiguities of life in zones of historic and contemporary conflict. As Hansen notes (and quotes from West),

> in a world where women are allowed to use their physical and mental capacities, "the young woman and the young man dash together out of adolescent into adult life like a couple of colts. But presently the woman looks around and sees that the man is not with her. He is some considerable distance behind her, not feeling very well." This depiction of men who cannot keep up but are left behind by women who are not even trying to beat them, but who simply enjoy the "pleasure" of using their minds and bodies, sits uneasily with [West's] construction of male strength and superiority and female weakness and submission. If women sacrifice their equality, this must be a strategic and unstable decision that cannot ultimately be sustained, regardless of whether "world peace" is achieved or not. (Pp. 124–125)

That is as good a statement on the grey zones of life and strategy as one finds – not because it is necessarily true but because it suggests an ambiguity about gender and war along with a certain undecidability about feminist pursuits.

If we were to ignore Hyun Jin's ongoing struggle with a past war and its abiding aftermath, we would miss another nugget of insight. It appears, impossibly, as though in dialogue with West's ambivalences about the Balkans and about gender truths that do not seem quite right:

> Sometimes I think I could have changed my story at this point in my life, just by choosing how to interpret what my father said. When he echoed his wife, repeating, "Blood will tell," I thought at the time that he was acknowledging that I could be nothing more than a whore [her biological mother is a sex worker]. But now I can almost believe he was reminding me that I was his daughter, that I carried his heart, that I had choices … Other times, I think the maps of our lives are etched into vein and muscle and bone, and that mere words – however interpreted – don't have the power to change anything. (Keller, 2002: 125–126)

Carolyn Nordstrom (2004: 10) might say that such mappings of survival when one's very being and daily circumstances are formed by war is part of a shadow network that exists at the intersection of "individual acts, national histories, and transnational cultures of militarization and economic gain." Within and between these grey zone networks are people's experiences with war and aftermaths that shape livelihoods, memories, social relations, and international relations. And one good place to spot them is in imaginative literatures and diaries of nonspecialist-specialists on war.

Museum zones of war

Another as yet unmentioned grey zone to enter as a site of war experience is the museum. I have been investigating of late the problematic of looting from museums during wars and other times of social upheaval (Sylvester, 2005; 2009). Looting is usually described as an economic activity that is awakened by opportunistic circumstance tied to greed. For Roger MacGinty (2004), there are also affective aspects of looting that can include efforts to enhance individual esteem by taking fine objects or the pleasure of owning and displaying them. MacGinty's interest in affective aspects of looting recalls Butler's (2004b) very different writing on the politics of grieving in wartime, and raises the question of whether some museum lootings carry meanings of grief, vulnerability and identity in situations where the built environment is being destroyed by air assaults or tanks are breaking apart Babylonian ruins. It would be naïve to think that greed and orders from metropolitan museums and auction-houses for specific pieces would not be at play in antiquities looting, especially since many archaeologists maintain that museums are the biggest and worst looters of heritage (e.g., Lunden, 2004). Yet war is a sensory experience for people and for objects (Sylvester, 2010b, 2011a). Those who physically touch antiquities connect the personal, the political, the sensory, and the power of the object in what could be termed a security scenario, allowing certain objects to live on as "companion[s] in life experience," as Sherry Turkle puts it in *Evocative Objects* (2007: 5); possibly these objects substitute for other things and identities lost in wars (Freud, 1991). It could also be the case that some antiquities are taken for safekeeping when their destruction could pose an existential threat to a community, as the linking of arms to protect the Egypt Museum from looters during the Arab Spring suggests.[6]

Anthropologist Nayanika Mookherjee (2011) has also shown interest in museums, objects, loss and commemorations of war around the liberation war that created Bangladesh out of East Pakistan in 1971. Her work focuses on the Bangladesh Liberation War Museum, which was established in 1996 with artifacts from ordinary Bangladeshi people, combatants and their families, and media depictions of the war and its victims. It is a museum that depicts war as experience and that focuses on displaying the experiences of everyday Bangladeshis while also linking that distinctive set of war experiences with museums of conscience internationally. The museum has chosen to emphasize and display very graphic incidents of violence that increasingly include the rapes that accompanied the war, estimated at 200,000. These displays build a case for an interpretation of the war as genocide, even though the conflict "is still termed as a civil war [with Pakistan] in the international legal language of human rights" (p. S85). To Mookherjee, the Bangladesh Liberation War Museum rests on an ambiguous instability that veers simultaneously toward remembrance and loss, and places the local in the global sphere of affective commemorations of the European Holocaust. The shadow of Giorgio Agamben (1998) appears, reminding us that the idea of genocidal exception has become part of global consciousness and periodic practice. We see some of that practice in institutions at intersections of war and arts.[7]

Other types of greyness signal incommensurability or incongruence as well as ambivalence: the war sufferer combatant, the feminist who wars for peace, the peaceful man assigned to the cruelly named demilitarized zone separating North from South Korea. Imagine that a war experience and story can change as a person remembers and narrates it, as the artist draws it or the journalist photographs it, as refugees escape it, as NGOs bring relief from it, or as one dons a masquerade of war by embellishing or telling untruths about personal war experiences. Grey zones comprise the everyday of international relations' war spectacles. They are full of truth and fiction and withholding, and they tend to be greyer than social science likes. Yet accepting and delving into dissonances and ambiguities, rather than trying to evade or fix them, is a key to unexpected insights. Think Methode, where fiction and fact, war and death strategy, and life within bare life mingle. Think of studying women of war and women against war together. Think about war re-enactments ethnographically. Pick the bones of war in cemeteries or mass graves, as Kwon has done. Think of a cyborg, whose agency in war is not wrapped up in a whole human body.

War questions for international relations and feminism

All academic fields have their flaws and their moments of brilliance. Feminist IR has begun to study war and experience, departing from the standard emphases of feminism and of IR. It is giving IR war studies flesh, depth, and stories about war that get far away from – and here let me say it directly – the hideous concept of collateral damage. Among the many, many questions they implicitly raise are these: collateral to what? Damage of what kind? One could work for years on such deceptively simple questions. And there are so many others. If war is conceptualized as a social institution, how does that institution work cross-nationally? That is, how does a social institution with multiple centers of authority operate across time, across cultures, and with differing political economies? How does it reproduce itself? This question seems tailor-made for mainstream constructivism, which has neglected war studies, and for critical IR devotees who study institutions in ways indebted to Foucault. More conventional approaches of international organization, global governance, and regime analysis would be useful to bring into this particular conversation, too.

For critical theorists, there is a parallel question of what war as phenomenological experience means, and how it operates. How might the changing identities and roles produced by one set of war experiences affect people in the future, and their memories of war and peace? Whose experiences and identities change in war and in which ways, and whose are less likely to change? I can imagine poststructuralist

IR becoming engaged by these issues as well as peace studies and European political philosophy.

What are the various bodies of war and how do they perform in different settings? What emotions fall under an umbrella concept of war and how do they differ or show similarities cross-culturally? What are some of the relays connecting body, mind, society, and the politics of war? These questions could align with feminist IR interests as well as with international political psychology and concerns with difference that animate both postcolonial studies and feminist IR.

Relatedly, to what extent are wars based in affect and emotion rather than rationality? Turn the question around: where and how does the rationality assumption prevalent across IR fit with notions of war as experience? I would be interested in hearing from the rational-choice theorists on that question as well as those who take a softer view of rationality, and others who refuse the rationality assumption. Again, linkage is the point, not dominance of one approach over others.

How do gender, race, generation, class, sexual preference and ethnicity intersect with various war experiences around the world? There is already a stable of feminist IR work, and some postcolonial analysis addressing "intersectionality" aspects of this question. It is a large topic, however, that international development studies might be well placed to probe as well. It is also important that feminist IR move beyond lengthy excursions into discussions of what gender is, and approach issues of war with the confidence that comes from knowing that feminist IR is an accepted knowledge camp within IR that has made notable contributions to the study of war.

Finally, I suggest that research on war and experience should draw from a wide range of literatures, fictional and factual, from social science and the arts and humanities, looking for insights, links, and for unexpected locations and types of war experience. The divisions between fields are artifacts of previous compartmentalizations of knowledge that might not make sense today. Rethinking "incommensurability" in research is called for, as are new ways for fields to reach toward one another for mutual identity, which Erin Manning's (2007) philosophical work suggests. At the same time there are already excellent examples of crossover work in areas of IR. In this regard, we recall Michael Shapiro's (2011) recent article on the domestic spaces and events of war-related politics and Der Derian's reach toward the military, technology, film, and the humanities as he writes on war.

For myself, the many war question(s) that derive from studying up from the ground, and seeking to join ideas pursued down from above, create collages of war as tools of war theorizing. Collage as an art-making technique creates and inhabits grey zones by juxtaposing unexpected things, leaving viewers to make sense of new combinations they might not have considered before. The collage is a visual challenge to the comforts of known connections, optical competence, laws of physics, design convictions, and common-sense relations. Collages disrupt usual story lines with their odd content and with the wrong materials for *haute art* that they are apt to use – house paints instead of oils, dirt from the bottom of one's shoe, coffee grounds, metal filings, old wallpaper. Art historian Brandon Taylor (2004: 19) says collages were once "outside the category of *beaux arts* representations, they were intrusive new entities with a hybrid ontology all their own."[8] They were once "a new category of thing" (p. 26). So is an IR orientation to war that takes experience as its key premise – experience with an enormously injurious institution. It is a new category of an old thing called war. It juxtaposes professional and amateur, sufferer and spectator, and erases the boundaries between them, just as collage began "to erase the line between professional and amateur, artist and artisan, that had largely underpinned the edifice of Western art since at least the Renaissance" (p. 28).

Taking war as experience has a similar collage effect of blurring high and low politics, places, and people. Butler aims for something similar in her feminist presentation of war as shared mourning that can build into an anti-war politics. It is a blurring of them, of enemy and friend, around the idea that war brings grief, no matter which side one is on, and us. One is reminded of John Berger's (1988) characterization of Picasso's famous painting, *Guernica*, as a depiction of war without any enemies to accuse. The moment captured by Picasso's painting is soul-wrenching, not because the cause behind the aerial bombardment of a Basque town during the Spanish Civil War was unjust, as it was, but because the attack terrifies those who experience it, no matter what the rationale or who carries it out.[9] Collage changes its own rules over time. The idea, however, that the contradictory, ambiguous, repetitive and unique experiences of war can be pasted side-by-side or on top of one another in ways that change all the original images, and show other sides of known phenomena, is fascinating as a model of war studies – for IR and for a pioneering feminist IR of war.

That brings us to the end of what is really a beginning. As I argued in *Art/Museums: International Relations Where We Least Expect It* (2009), IR is at an end and art is at an end in the sense that both fields of theory have been outpaced by events and practices. IR goes about its business in true camp spirit today but is hard-pressed to bring its considerable accumulated knowledge to bear on a troubled warring world, where bombs sit in theaters or on the bus, art empties into the streets of war, and women can be raped with abandon in war zones. IR has not prepared itself for the range of experiences that so many can access on 24/7 news stations and YouTube. Fortunately, feminist IR is making a strong start in war studies; and art production carries on as though there were theories that could explain its ways and means today. This book could cover only a sliver of the research and presentations that have dealt with war – apologies to all who have read this far hoping to see a glimpse of yourself. Space has limits, but the study of war that features flesh and blood, cyborg and robotic and performing bodies, emotions, and social relations is expanding. Not a moment too soon, either – for all involved.

NOTES

Introduction: War questions for feminism and International Relations

1 For a recent account of debates on war in the realm of civil war studies alone, see e.g., Mundy (2011).
2 The last werewolf in the novel of the same name (Duncan, 2011) is 400 years old.
3 There has been some interest in the experiences of diplomats (Adler-Nissen, 2008; Neumann, 2008) and issues of practicality in everyday international relations (Pouliot, 2008).
4 Spelman does go on to clarify her own view, quoting René Descartes in *Meditations VI*: "All these sensations of hunger, thirst, pain, etc. are in truth none other than certain confused modes of thought which are produced by the union and apparent intermingling of mind and body" (in Wilson, 1969: 216, quoted in Spelman, 1997: 173).
5 Isak Svensson and Mathilda Lindgren (2011) report a rise in nonviolent uprisings in East Asia since 1979, as well.
6 Personal correspondence. Hansen adds that the controversies in security studies over how and whether the traditional concept of national security should be expanded can be seen as a debate over how close or remote the possibility of force must be for something to qualify as "security."
7 I thank Lene Hansen for suggesting this point in personal correspondence.

1 IR takes on war

1 For a balanced review of Mearsheimer's and Walt's work on the Israel lobby, and those whose replies support or critique it, see Slater (2009).
2 From comments made to the Experiencing War workshop held at the School of Global Studies, University of Gothenburg, April 14–15, 2011. See her *International Authority and the Responsibility to Protect* (2011).
3 James Peck (2010) argues that human rights discourse and practices have long been systematically subverted, redefined, or sidetracked by realist American administrations.
4 Chan made these specific critiques at an Experiencing War workshop held at Gothenburg University, April 2011, and at a round-table discussion on that project and the book, *Experiencing War* (Sylvester, 2011a), sponsored by IDEAS at the London School of Economics, January 2011.

5 Some of Fierke's other work (e.g., 1996, 1998) fits this description and can be ranged under critical constructivism, especially her studies of language games.
6 Another anti-war group of women, Code Pink, conflates the private and the public realms of "femininity" into a powerful public anti-war strategy. For a participant-observational-based study of Code Pink, see Rowe (2013).
7 Lauren Berlant (2004: 10) suggests there is a similar training in aversion to emotions like compassion, a point to which we return in Chapter 4 while discussing the body, emotion, and war.

2 Feminist (IR) takes on war

1 Eric Blanchard (2011) asks where gender is in the popular British theoretical tradition called the English School. That school of thought famously comes up with a world society view of international relations that emphasizes cooperative cosmopolitanism over realist brute force, but does so without discussing feminist IR literature on cooperation.
2 Harry Windsor, grandson of Britain's Queen Elizabeth, served briefly in the Afghanistan war and expressed tremendous disappointment when his presence there became known and he was evacuated back to the UK. It is the kind of tale that presents war as a young man's dream, even if that young man has a life outside war that could easily protect him from military service.
3 There are other important works on this subject at the edges of IR and other fields, including Stiehm (1984), Cooke and Woollacott (1993), and Hearn (1998).
4 Rajeev Patel and Philip McMichael (2004: 236) have a very direct way of putting this: the inheritors of the state see themselves "as sinned against and unsinning." They correctly demonize the imperial apparatus of control "without implicating themselves in its functioning." See discussion in Sylvester (2006).
5 Jill Gibbon (2011) offers an unusual view of arms fairs through the many drawings she has made, clandestinely, at several London arms fairs.
6 E.g. seeing tanks unexpectedly bearing down on him during his field observations of war games, Der Derian (2009: 8) says: "I froze, feeling that terrifying yet seductive rush that comes when the usual boundaries, between past and present, war and game, spectator and participant break down ... Detached and yet connected to a dangerous situation by a kind of traumatic voyeurism, I watched myself watch the tanks bear down. In this moment elongated by terror, I entered the borderlands of simulation, where fear and fun, friend and foe, all blur together." Alison says at one point about her interviews in Northern Ireland: "my richer awareness of all potentialities meant that I felt more afraid (occasionally) ... than I did in Sri Lanka ... I had absorbed the baseline fear and constant watchfulness of Northern Irish people" (p. 30).

3 War as physical experience

1 Despite its flaws, "new wars" analysis does recognize that many of today's wars purposely target civilians (see Chapter 1). It is also possible to see war as so socially embedded in certain new wars zones, and so sacrificing of life itself, that the state of war becomes a practice of the self in those areas (Mbembe, 2002: 269).
2 These are the kinds of monsters Lady Gaga calls her fans and encourages everyone to be and embrace; a music video shows her being born as a composite, genderless creature, something she celebrates in her song 'Born That Way'. Pink's lyrics also advocate inclusiveness, nondiscrimination, and acceptance of "nerds" ("too school for cool") in her song 'Raise Your Glass'.
3 Citing space limitations in a pocket-sized paperback only 150 pages long, Drezner offers one sort-of dig and one sort-of compliment to feminism: "Marxists and feminists would likely sympathize more with the zombies [than with those eaten by them]" (p. 17); yet he

notes a sharp contrast in the *Dawn of the Dead* film of 2004 between the way the security guards react to zombies invading the mall and "when the female lead (a nurse) and her compatriots are in charge. Decision making is both more consensual and yields superior results under the latter regime" (p. 105). Stereotyping?

4 For additional insights into women's war experiences in the Balkans, see e.g., Chapter 9 in Hansen's (2006) discourse analytic treatment of the Balkan wars, where she discusses gendered genocide in Bosnia; Janie L. Leatherman's (2010: 29) constructivist analysis of sexual violence as "a runaway norm that justifies and normalizes extreme forms of violence ...''; and Inger Skjelsbaek's (2012) interview-based study of the psychology of war rape in Bosnia and Herzegovina.

5 From personal conversations with the researchers.

6 Utas discussed this problem at the Experiencing War workshop held at Gothenburg University, April 2011.

7 This is a question that the anonymous writer of *A Woman in Berlin* (2000: 63) also asks in the context of the daily multiple rapes she experiences as a 34-year-old single woman living during the Russian "liberation" of Berlin in 1945. She writes of the appalling conditions in the city at the end of the war, of standing in long lines in the hope of getting some rancid butter while rockets periodically fall, perhaps killing a person in the middle of the queue; everyone pushes the barely dead body aside and continues their butter vigil. In that situation, where horrors pile on top of horrors, she asks: "What does it mean − rape? When I said the word for the first time aloud, Friday evening in the [bomb shelter] basement, it sent shivers down my spine. Now I can think it and write it with an untrembling hand, say it out loud to get used to hearing it said. It sounds like the absolute worst, the end of everything − but it's not."

4 War as emotional experience

1 Hannah Arendt (1989) famously insisted that emotions are even anti-political.

2 For a brave and honest assessment of feminist researcher fear, see Elina Penttinen's (2007) discussion of sitting alone in a parked car during a blizzard, waiting to witness the arrival of Russian women and their handlers along an alleged trafficking highway in northern Finland. Far less tense but no less emotional were moments in my interviews of owners and managers of Zimbabwean businesses that employed women, as well as government officials supposedly committed to helping women workers improve their status, when I could not believe what I was hearing and actually challenged the speaker (Sylvester, 2000, 2011c).

3 The workshop, titled Women: Armed and Dangerous?, was held March 2009 as part of research activities I directed at Lancaster University called Touching War.

4 The literature on emotions as studied within IR and in psychology is now vast, far larger than any short review could indicate. Crawford (2000) provides an excellent review of some of that literature, as does Mercer (2005).

5 One difference of view is articulated by William Long and Peter Becker (2003: 125), who state that neuroscience has proven that "[f]eelings alert us to problems our emotions have already begun to solve," whereas others refer to intensities (presumably bodily feelings) that precede emotions as social components, as Massumi (2002) puts this.

6 Rational choice theorists also assume that micropolitical aspects, understood as individual-based behaviors, are important, but characterizes these differently than neuropolitics analysts would. They speak of individual constraints on rationality, of calculations, and of deductive reasoning as micropolitical. Emotions are separated from rationality and are deemed irrelevant to rational decision-making in all individuals. See Bueno de Mesquita (2010). William Long and Peter Becker (2003: 121) describe rational choice as "folk psychology. That is, it assumes, but does not examine, that the mind is a general-purpose, universally logical, problem-solving mechanism."

7 From personal conversations with the researchers.

8   Butler does refer to Berlant's ideas in *Frames of War* (2010: 34), where she takes up the affect studies approach and argues that "a certain interpretive act implicitly takes hold at moments of primary affective responsiveness." Butler's inclination to separate feelings from emotion comes across as forced after her more raw and forceful political arguments in *Precarious Life*. In *Frames of War*, grief and mourning seem like lumps of coal awaiting a match.

5 Concluding, collaging, and looking ahead

1   Later in the article, Smith does suggest that the gist of his argument is aimed at IR's mainstream, especially its rational choice contingent (p. 504).
2   This is a point Lipschutz made in a seminar at the Department of Political Science, Lund University, November 9, 2011.
3   The concept of grey zones is often associated with meaningless choices that can lead to evil outcomes or to trauma, often in the context of the Holocaust or of feminist ethics. See Card (1999), Levi (1987), and Schott (2011). I am obviously using the term in a broader sense.
4   There is some concern that Scarry's view is tied to an implicit dualism of body–mind and self–other (Lee, 2005). Yet Das's perspective, while appealing to social science, resonates with affect studies that can overemphasize the social components of all body-based reactions.
5   Hansen's concern to read this travelogue-cum-political treatise as a source for IR comes out of remarks Vivienne Jabri and I made at a round-table discussion in 2007 at the London School of Economics. It was on "Reflections on the Past, Prospects for the Future in Gender and International Relations" (*Millennium*, 2008; also Sylvester, 2010a) and the remarks called for greater attention to classical texts (Jabri) and to feminists whose views were difficult to embrace.
6   Of course, some looters vandalize museums and damage or destroy antiquities on their rampages, a subject that has also not been treated adequately in the literature.
7   IR discussions of museum intersections with aspects of war experience include Lisle (2006) and Chapters 2 ("Nuclear Reactions: The (Re) Presentation of Hiroshima at the National Air and Space Museum") and 3 ("Memorializing Mass Murder: The United States Holocaust Memorial Museum") in Luke (2002).
8   Many of Picasso's paper collages of 1912 and 1913 refer in words or images to the military conflict in the Balkans, which broke out in October 1912, as do works by other early collagists, principally the Futurists Gino Severini and Filippo Martini, who embedded stories and photographs of World War I trench warfare in their works.
9   See Ann Orford's (2011) discussion of *Guernica* and the lengths to which the UN Security Council went to mask a tapestry rendition of it before then Secretary of State Colin Powell arrived to present the American rationale for invading Iraq, 2003.

REFERENCES

Abramson, Rita and Michael C. Williams (2010), *Security Beyond the State: Private Security in International Politics* (Cambridge: Cambridge University Press).

Ackerly, Brooke, Maria Stern, and Jacqui True, eds. (2006), *Feminist Methodologies for International Relations* (Cambridge: Cambridge University Press).

Adler-Nissen, Rebecca (2008), "The Diplomacy of Opting Out: A Bourdieudian Approach to National Integration Strategies," *Journal of Common Market Studies*, 46, 3: 663–684.

Afshar, Haleh and Deborah Eade, eds. (2003), *Development, Women, and War: Feminist Perspectives* (Bloomfield, IN: Kumarian Press).

Agamben, Giorgio (1998), *Homo Sacer, Sovereign Power and Bare Life* (Stanford, CA: Stanford University Press).

Ahmed, Sara (2000), *Strange Encounters: Embodied Others in Post-Coloniality* (London: Routledge).

Ahmed, Sara (2004), "Affective Economies," *Social Text*, 22, 2: 117–139.

Alexander, Karen and Mary Hawkesworth, eds. (2008), *War and Terrorism: Feminist Perspectives* (Chicago: University of Chicago Press).

Al-Ali, Nadje Sadig and Nicola Pratt, eds. (2009), *Women and War in the Middle East: Transnational Perspectives* (London: Zed Press).

Ali, Ayaan Hirsi (2006), *The Caged Virgin: A Muslim Woman's Cry for Reason* (New York: Free Press).

Ali, Ayaan Hirsi (2007), *Infidel: My Life* (New York: Free Press).

Ali Ayaan Hirsi (2010), *Nomad: From Islam to America* (New York: Simon and Schuster).

Alison, Miranda (2009), *Women and Violence: Female Combatants in Ethno-National Conflict* (London: Routledge).

Allison, Graham (1969), "Conceptual Models and the Cuban Missile Crisis," *American Political Science Review*, 63, 3: 689–718.

Anonymous (2000), *A Woman in Berlin: Eight Weeks in the Conquered City. A Diary*, Philip Boehm, trans. (New York: Metropolitan Books/Henry Holt and Company).

Aradau, Claudia (2004), "The Perverse Politics of Four-Letter Words: Risk and Pity in the Securitization of Human Trafficking," *Millennium: Journal of International Studies*, 33, 2: 251–277.

Arendt, Hannah (1958), *The Human Condition* (Chicago: University of Chicago Press).

Baaz, Maria Eriksson and Maria Stern (2008), "Making Sense of Violence: Voices of Soldiers in the Congo (DRC)," *Journal of Modern African Studies*, 46, 1: 57–86.

Baaz, Maria Eriksson and Maria Stern (2009), "Why Do Soldiers Rape? Masculinity, Violence, and Sexuality in the Armed Forces in the Congo," *International Studies Quarterly*, 53: 495–518.

Baaz, Maria Eriksson and Maria Stern (2010), *The Complexity of Violence: A Critical Analysis of Sexual Violence in the Democratic Republic of Congo* (Stockholm: SIDA).

Bal, Mieke (1997), *Narratology: Introduction to the Theory of Narrative*, 2nd ed. (Toronto: University of Toronto Press).

Balzacq, Thierry and Robert Jervis (2004), "Logics of Mind and International System: A Journey with Robert Jervis," *Review of International Studies*, 30: 564–565.

Barkawi, Tarak (2004a), "On the Pedagogy of 'Small Wars,'" *International Affairs*, 80, 1: 19–38.

Barkawi, Tarak, (2004b), "Connection and Constitution: Locating War in Globalization Studies," *Globalizations*, 1, 2: 155–170.

Barkawi, Tarak (2006), *Globalization and War* (Lanham, MD: Rowman and Littlefield).

Barkawi, Tarak (2011), "From War to Security: Security Studies, the Wider Agenda and the Fate of the Study of War," *Millennium: Journal of International Studies*, online 22 March. http://mil.sagepub.com/content/early/2011/03/09/0305829811400656.

Barkawi, Tarak and Shane Brighton (2011), "Powers of War: Fighting, Knowledge, and Critique," *International Political Sociology*, 5, 2: 126–143.

Bennett, Jane (2010), *Vibrant Matter: A Political Ecology of Things* (Durham, NC: Duke University Press).

Berger, John (1988), "Success and Failure of Picasso," in Ellen Oppler, ed., *Picasso's* Guernica (New York: W.W. Norton).

Berlant, Lauren (1993), "The Queen of America Goes to Washington City: Harriet Jacobs, Frances Harper, Anita Hill," *American Literature*, 65, 3: 549–574.

Berlant, Lauren (2000), "The Subject of True Feeling: Pain, Privacy, and Politics," in Sara Ahmed, ed., *Transformations: Thinking Through Feminism* (London: Routledge): 33–48.

Berlant, Lauren (2004), "Introduction: Compassion (and Withholding)," in Lauren Berlant, ed., *Compassion: The Culture and Politics of an Emotion* (New York: Routledge): 1–13.

Biccum, April (2005), "Development and the 'New' Imperialism: A Reinvention of Colonial Discourse in DFID Promotional Literature," *Third World Quarterly*, 26: 1005–1020.

Bishop, Kyle (2008), "The Sub-Altern Monster: Imperialist Hegemony and the Cinematic Voodoo Zombie," *Journal of American Culture*, 31, 2: 141–152.

Blanchard, Eric (2011), "Why is There No Gender in the English School?," *Review of International Studies*, 37, 2: 855–879.

Bleiker, Roland and Emma Hutchison (2008), "Fear No More: Emotions and World Politics," *Review of International Studies*, 34, 115–135.

Bordo, Susan (2003), *Unbearable Weight: Feminism, Western Culture, and the Body* (Berkeley: University of California Press).

Braidotti, Rosi (1989), "The Politics of Ontological Difference," in Teresa Brennan, ed., *Between Feminism and Psychoanalysis* (London: Routledge): 89–105.

Brighton, Shane (2011), "Three Propositions on the Phenomenology of War," *International Political Sociology*, 5, 1: 101–105.

Brown, Wendy (1995), *States of Injury: Power and Freedom in Late Modernity* (Princeton: Princeton University Press).

Bueno de Mesquita (2010), *The Predictioneer's Game: Using Brazen Self-Interest to See and Shape the Future* (New York: Ramdom House).

Burke, Anthony (2007), *Beyond Security, Ethics and Violence: War Against the Other* (London: Routledge).

Butler, Judith (1993), *Bodies That Matter: On the Discursive Limits of "Sex"* (New York: Routledge).

Butler, Judith (2004a), *Undoing Gender* (New York: Routledge).

Butler, Judith (2004b), *Precarious Life: The Powers of Mourning and Violence* (London: Verso).

Butler, Judith (2010), *Frames of War: When is Life Grievable?* (London: Verso).

Buzan, Barry and Lene Hansen (2009), *The Evolution of International Security Studies* (Cambridge: Cambridge University Press).

Caprioli, Mary (2003), "Gender Equality and State Aggression: The Impact of Domestic Gender Equality on State First Use of Force," *International Interactions*, 29, 3: 195–214.

Caputo, Philip (1977), *A Rumor of War* (London: Macmillan).

Card, Claudia (1999), "Groping through Gray Zones," in Claudia Card, ed., *On Feminist Ethics and Politics* (Lawrence: University Press of Kansas): 3–26.

Carpenter, Charli (2003), "Women and Children First: Gender, Norms, and Humanitarian Evacuation in the Balkans, 1991–1995," *International Organization*, 57, 4: 661–694.

C.A.S.E. Collective (2006), "Critical Approaches to Security in Europe: A Networked Manifesto," *Security Dialogue* 37, 4: 443–487.

Chan, Stephen (2011), "On the Uselessness of New Wars Theory: Lessons from African Conflicts," in Christine Sylvester, ed., *Experiencing War* (London: Routledge): 94–102.

Cockburn, Cynthia (2007), *From Where We Stand: War, Women's Activism, and Feminist Analysis* (London: Zed).

Cohn, Carol (1987), "Sex and Death in the Rational World of Defense Intellectuals," *Signs: Journal of Women in Culture and Society*, 14, 4: 687–718.

Comaroff, Jean and John Comaroff (2002), "Alien-Nation: Zombies, Immigrants, and Millennial Capitalism," *South Atlantic Quarterly*, 101 (Fall): 779–805.

Connolly, William (2002), *Neuropolitics: Thinking, Culture, Speed* (Minneapolis: University of Minnesota Press).

Connolly, William (2011), "Critical Response: The Complexity of Intention," *Critical Inquiry*, 37, 4: 791–798.

Cooke, Miriam and Angela Woollacott, eds. (1993), *Gendering War Talk* (Princeton: Princeton University Press).

Courtemanche, Gil (2003), *A Sunday at the Pool in Kigali* (London: Canongate).

Crawford, Neta (2000), "The Passion of World Politics: Propositions on Emotion and Emotional Relationships," *International Security*, 24, 2: 116–156.

Daly, Mary (1978). *Gyn/Ecology: The Metaethics of Radical Feminism* (Boston: Beacon Press).

Damasio, Anthony (1994), *Descartes' Error: Emotion, Reason, and the Human Brain* (New York: Avon).

Damasio, Anthony (2000), *The Feeling of What Happens: Body and Emotion in the Making of Consciousness* (New York: Harcourt Brace).

Das, Veena (1995), *Critical Events: An Anthropological Perspective on Contemporary India* (New Delhi: Oxford University Press).

Das, Veena (2007), *Life and Words: Violence and the Descent into the Ordinary* (Berkeley: University of California Press).

Davies, Matt (2010), "'You Can't Charge Innocent People for Saving Their Lives!' Work in 'Buffy the Vampire Slayer,'" *International Political Sociology*, 4, 2: 178–95.

Davis, Angela (1989), *Angela Davis: An Autobiography* (New York: International Publications).

D'Costa, Bina (2006), "Marginalized Identity: New Frontiers of Research for IR?," in Brooke Ackerly, Maria Stern and Jacqui True, eds., *Feminist Methodologies for International Relations* (Cambridge: Cambridge University Press), 129–152.

De Beauvoir, Simone (1957), *The Second Sex* (New York: Alfred A. Knopf).

Der Derian, James (2000), "Virtuous War/Virtuous Theory," *International Affairs*, 76, 4: 771–788.

Der Derian, James (2005), "Imaging Terror: Logos, Pathos, Ethos," *Third World Quarterly*, 26, 1: 23–37.

Der Derian, James (2009), *Virtuous War: Mapping the Military-Industrial Media-Entertainment Network*, 2nd edition (New York: Routledge).

Der Derian, James, David Udris, and Michael Udris (2009), *Human Terrain*, a film.

Deudney, Daniel and John Ikenberry (1991/92), "Soviet Reform and the End of the Cold War," *Review of International Studies*, 17, 3: 225–250.

Dillon, Mick and Julian Reid (2001), "Global Liberal Governance: Biopolitics, Security, and War," *Millennium: Journal of International Studies*, 30, 1: 41–66.

Drezner, Daniel (2011), *Theories of International Politics and Zombies* (Princeton: Princeton University Press).

Duffield, Mark (2001), *Global Governance and the New Wars: The Merging of Development and Security* (London: Zed).

Duffield, Mark (2007), *Development, Security and Unending War* (Cambridge: Polity).

Duncan, Glen (2011), *The Last Werewolf* (New York: Knopf).

Edkins, Jenny (2003), *Trauma and the Memory of Politics* (Cambridge: Cambridge University Press).

Elshtain, Jean Bethke (1987), *Women and War* (New York: Basic Books).

Elshtain, Jean Bethke (1988), "The Problem with Peace," *Millennium: Journal of International Studies*, 17 (3): 441–449.

Elshtain, Jean Bethke (2005), "Against the New Utopianism," *Ethics and International Affairs*, 19, 2: 91–95.

Elshtain, Jean Bethke (2009), "Woman, the State, and War," *International Relations* 23, 2: 289–303.

Enloe, Cynthia (1983), *Does Khaki Become You? The Militarization of Women's Lives* (London: Pandora).

Enloe, Cynthia (1989), *Bananas, Beaches, and Bases: Making Feminist Sense of International Relations* (London: Pandora).

Enloe, Cynthia (1993), *The Morning After: Sexual Politics at the End of the Cold War* (Berkeley: University of California Press).

Enloe, Cynthia (2000), *Maneuvers: The International Politics of Militarizing Women's Lives* (Berkeley: University of California Press).

Enloe, Cynthia (2004), *The Curious Feminist* (Berkeley: University of California Press).

Enloe, Cynthia (2010), *Nimo's War, Emma's War: Making Feminist Sense of the Iraq War* (Berkeley: University of California Press).

Fanon, Frantz (1963), *The Wretched of the Earth* (New York: Grove Press).

Fay, Jennifer (2008), "Dead Subjectivity: White Zombie, Black Baghdad," *CR: The New Centennial Review*, 8 (Spring): 81–101.

Fearon, James (1995), "Rationalist Explanations for War," *International Organization*, 49, 3: 379–414.

Ferguson, Kathy (1993), *The Man Question: Visions of Subjectivity in Feminist Theory* (Berkeley: University of California Press).

Fierke, Karin (1996), "Multiple Identities, Interfacing Games: The Social Construction of Western Action in Bosnia," *European Journal of International Relations*, 2, 4: 467–497.

Fierke, Karin (1998), *Changing Games, Changing Strategies: Critical Investigations in Security* (Manchester: University of Manchester Press).

Fierke, Karin (2004), "Whereof We Can Speak, Thereof We Must Not Be Silent: Trauma, Political Solipsism and War," *Review of International Studies*, 30: 471–491.

Finnemore, Martha (2003), *The Purpose of Intervention: Changing Beliefs about the Use of Force* (Ithaca, NY: Cornell University Press).

Finnemore, Martha and Kathryn Sikkink (1998), "International Norm Dynamics and Political Change," *International Organization*, 52, 2: 887–917.

Finnemore, Martha and Kathryn Sikkink (2001), "Taking Stock: The Constructivist Research Program in International Relations and Comparative Politics," *Annual Review of Political Science*, 4: 391–416.

Freud, Sigmund (1991), *On Metapsychology: The Theory of Psychoanalysis* (London: Penguin).

Friedan, Betty (1963), *The Feminine Mystique* (New York: W.W. Norton).

Fukuyama, Francis (1989), "The End of History?" *National Interest*, 16 (September): 3–18.

Gerecke, Megan (2010), "Explaining Sexual Violence in Conflict Situations," in Laura Sjoberg and Sandra Via, eds., *Gender, War, and Militarism: Feminist Perspectives* (New York: Praeger): 138–154.

Gibbon, Jill (2011), "Dilemmas of Drawing War," in Christine Sylvester, ed., *Experiencing War* (London: Routledge): 103–117.

Gleditsch, N.P., Peter Wallensteen, Maria Eriksson, M. Sollenberg, and Strand, H. (2002), "Armed Conflict 1946–2001: A New Data Set," *Journal of Peace Research*, 39, 5: 615–637.

Gleditsch, N. (2008), "The Liberal Moment Fifteen Years on," *International Studies Quarterly*, 22, 4: 691–712.

Gobodo-Madikizela, Pumla (2008), "Empathetic Repair After Mass Trauma When Vengeance is Arrested," *European Journal of Social Theory*, 11, 3: 331–350.

Goldstein, Joshua (2001), *War and Gender* (Cambridge: Cambridge University Press).

Goldstein, Joshua (2011), *Winning the War on War: The Decline of Armed Conflict Worldwide* (New York: Dutton Adult).

Grant, Judith (1993), *Fundamental Feminisms: Contesting the Core Concepts of Feminist Theory* (New York: Routledge).

Gregg Melissa, Seigworth and Gregory and eds. (2010), "An Inventory of Shimmers," in *The Affect Theory Reader* (Durham, NC: Duke University Press): 1–25.

Gronemeyer, Marianne (1992), "Help," in Wolfgang Sachs, ed., *The Development Dictionary: A Guide to Knowledge as Power* (London: Zed Books).

Grossberg, Lawrence (2010), "Affect's Future: Rediscovering the Virtual in the Actual," in Melissa Gregg and Gregory Seigworth, eds., *The Affect Theory Reader* (Durham, NC: Duke University Press): 309–338.

Grosz, Elizabeth (1994), *Volatile Bodies: Towards a Corporeal Feminism* (Bloomington, IN: Indiana University Press).

Hallinan, Joseph T. (2011), "The Young and the Perceptive," *New York Times*, Week in Review, March 6: 10.

Hansen, Lene (2006), *Security as Practice: Discourse Analysis and the Bosnian War* (London: Routledge).

Hansen, Lene (2011), "A Research Agenda on Feminist Texts and the Gendered Constitution of International Politics in Rebecca West's *Black Lamb and Grey Falcon*," *Millennium: Journal of International Studies*, 40, 1: 109–128.

Harari, Yuval Noah (2008) "Combat Flow: Military, Political, and Ethical Dimensions of Subjective Well-Being in War", *Review of General Psychology* 12: 253–264.

Haraway, Donna (1990), "A Manifesto for Cyborgs: Science, Technology, and Socialist Feminism in the 1980s," in Linda Nicholson, ed., *Feminism/Postmodernism* (New York: Routledge): 190–233.

Haraway, Donna (1991), "The Biopolitics of Postmodern Bodies: Constitutions of Self in Immune System Discourse," in *Simians, Cyborgs, and Women: The Reinvention of Nature* (New York: Routledge): 203–230.

Haraway, Donna (1997), *Modest_Witness @ Second_Millennium. FemaleMan_Meets_OncoMouse* (New York: Routledge).

Harding, Sandra (1986), *The Science Question in Feminism* (Ithaca, NY: Cornell University Press).

Harding Sandra (1991), *Whose Science? Whose Knowledge? Thinking from Women's Lives* (Ithaca, NY: Cornell University Press).

Hearn, Jeff (1998), *The Violences of Men* (London: Sage).

Herz, John (2003), "The Security Dilemma in International Relations: Background and Present Problems," *International Relations*, 17: 411–416.

Higate, Paul (ed.) (2003), *Military Masculinities: Identity and the State*. (Westport, CT: Praeger).

Hill, Christopher (2003), *The Changing Politics of Foreign Policy* (Houndmills: Palgrave).

Holzner, Brigitte (2011), "Wars, Bodies, and Development," in Christine Sylvester, ed., *Experiencing War* (London: Routledge): 42–63.

hooks, bell (1990), *Yearning: Race, Gender, and Cultural Politics* (Boston: South End Press).

Hove, Chenjerai (1994), *Shebeen Tales: Messages from Harare* (Harare: Baobab Books).

Hudson, Valerie, Mary Caprioli, Bonnie Ballif-Spanville, Rose McDermott, and Chad Emmett (2009), "The Heart of the Matter: The Security of Women and the Security of States," *International Security*, 33, 3: 7–45.

Huntington, Samuel (1993), "The Clash of Civilizations," *Foreign Affairs*, 73, 3: 22–49.

Hutchings, Kimberly (2011), "Gendered Humanitarianism: Reconsidering the Ethics of War," in Christine Sylvester, ed., *Experiencing War* (London: Routledge): 28–41.

Huysmans, Jef (1998), "Security! What Do You Mean? From Concept to Thick Signifier," *European Journal of International Relations*, 4, 2: 226–255.

Irigaray, Luce (1985), *This Sex Which is Not One* (Ithaca, NY: Cornell University Press).

Jabri,Vivienne (2010), *War and the Transformation of Global Politics,* 2nd ed. (New York: Palgrave Macmillan).

Jacoby, Tami (2006), "From the Trenches: Dilemmas of Feminist IR Fieldwork," in Brooke Ackerly, Maria Stern and Jacqui True, eds., *Feminist Methodologies for International Relations* (Cambridge: Cambridge University Press), 153–173.

James, William (1976), *Essays in Radical Empiricism*, ed. Frederick Burkhardt *et al.* (Cambridge, MA: Harvard University Press).

Kaldor, Mary (1998), New and Old Wars (Stanford, CA: Stanford University Press).

Kaldor, Mary (2006), *New and Old Wars: Organised Violence in a Global Era*, 2nd edition (Stanford, CA: Stanford University Press).

Kapoor, Ilan (2008), *The Postcolonial Politics of Development* (New York: Routledge).

Keller, Nora Okja (2002), *Fox Girl* (New York: Penguin Books).

Keohane, Robert (1990), "Empathy and International Regimes," in Jane Mansbach, ed., *Beyond Self-Interest* (Chicago: University of Chicago Press): 227–236.

Kidder, Tracy (2010), *Strength in What Remains* (New York: Random House Paperback Edition).

Kinnvall, Catarina and Paul Nesbitt-Larking (2009), "Security, Subjectivity and Space in Postcolonial Europe: Muslims in the Diaspora," *European Security*, 18, 3: 305–325.

Kinnvall, Catarina and Jitka Linden (2010), "Dialogical Selves Between Security and Insecurity: Migration, Multiculturalism, and the Challenge of the Global," *Theory and Psychology*, 20, 5: 595–619.

Kronsell, Annica (2006), "Methods for Studying Silences: Gender Analysis in Institutions of Hegemonic Masculinity," in Brooke Ackerly, Maria Stern, and Jacqui True, eds., *Feminist Methodologies for International Relations* (Cambridge: Cambridge University Press): 108–128.

Kronsell, Annica (2012), *Gender, Sex and the Post-National Defense: Militarism and Peacekeeping* (Oxford: Oxford University Press).

Kronsell, Annica and Erika Svedberg, eds. (2012), *Making Gender, Making War: Violence, Military, and Peacekeeping Practices* (London: Routledge).

Kuehnast, Kathleen, Chantal de Jonge Oudraat, and Helga Hernes, eds. (2011), *Women and War: Power and Protection in the 21st Century* (Washington: The United States Institute of Peace Press).

Kumar, Krishna, ed. (2001), *Women and Civil War: Impact, Organizations, and Action* (Boulder, CO: Lynne Rienner).

Kwon, Heonik (2010), *The Other Cold War* (New York: Columbia University Press).

Kwon, Heonik (2011), "Experiencing the Cold War," in Christine Sylvester, ed., *Experiencing War* (London: Routledge): 79–93.

Lasswell, Harold (1965), *World Politics and Personal Insecurity* (New York: Free Press).

Leander, Anna, ed. (2012), *Commercializing Security in Europe: Political Consequences for Peace and Reconciliation Operations* (London: Routledge).

Leatherman, Janie L. (2011), *Sexual Violence and Armed Conflict* (Cambridge: Polity Press).

Lebow, Richard Ned (2005), "Reason, Emotion and Cooperation," *International Politics*, 42: 283–313.

Lebow, Richard Ned (2010), *Why Nations Fight* (Cambridge: Cambridge University Press).

Lebow, Richard Ned and Thomas Risse-Kappan, eds. (1997), *International Relations Theory and the End of the Cold War* (New York: Columbia University Press).

Lee, Wendy (2005), "On the (Im)materiality of Violence," *Feminist Theory* 6, no. 3: 277–295.

Levi, Primo (1987), *If This is a Man* and *The Truce* (New York: Vintage).

Levinas, Emmanuel (1969), "Totality and Infinity: An Essay on Exteriority" (Pittsburgh, PA: Duquesne University Press).

Levy, Jack and William R. Thompson (2010), *Causes of War* (London: Wiley-Blackwell).

Leys, Ruth (2011), "The Return to Affect: A Critique," *Critical Inquiry*, 37, 3: 434–472.

Lisle, Debbie (2006), "Sublime Lessons: Education and Ambivalence in War Exhibitions," *Millennium: Journal of International Studies*, 34, 3: 841–862.

Lloyd, Anthony (1999), *My War Gone By, I Miss It So* (London: Anchor).

Long, William and Peter Becker (2003), *War and Reconciliation: Reason and Emotion in Conflict Resolution* (Cambridge: MIT Press).

Luke, Timothy (2002), *Museum Politics: Power Plays at the Exhibition* (Minneapolis: University of Minnesota Press).

Lunden, Staffan (2004), "The Scholar and the Market: Swedish Scholarly Contributions to the Destruction of the World's Archaeological Heritage," in Håkan Karlsson, ed., *Swedish Archaeologists on Ethics* (Lindome: Bricoleur Press): 197–247.

MacGinty, Roger (2004), "Looting in the Context of Violent Conflict: A Conceptualization and Typology," *Third World Quarterly*, 25, 5: 857–870.

MacKenzie, Megan (2009), "Securitization and Desecuritization: Female Soldiers and the Reconstruction of Women in Post-Conflict Sierra Leone," *Security Studies*, 18, 2: 241–261.

MacKenzie, Megan (2011a), "Ruling Exceptions: Female Soldiers and Everyday Experiences of Civil Conflict," in Christine Sylvester, ed., *Experiencing War* (London: Routledge): 64–78.

MacKenzie, Megan (2011b), "Their Personal is Political, Not Mine: Feminism and Emotion," in Christine Sylvester, ed., "Forum: Emotion and the Feminist IR Researcher," *International Studies Review*, 13, 4: 691–693.

MacKenzie, Megan (forthcoming, 2013), *Female Soldiers in Sierra Leone: Sex, Security, and Post–Conflict Development* (New York: New York University Press).

Malesevic, Sinisa (2010), *The Sociology of War and Violence* (Cambridge: Cambridge University Press).

Manning, Erin (2007), *Politics of Touch: Sense, Movement, Sovereignty* (Minneapolis, MN: University of Minnesota Press).

Marshall, Sandra (2011), "Super-Human Researchers in Feminist International Relations' Narratives," *International Studies Review*, 13: 1–22.

Massumi, Brian (2002), *Parables for the Virtual: Movement, Affect, Sensation* (Durham, NC: Duke University Press).

Massumi, Brian (2010), "The Future Birth of the Affective Fact: The Political Ontology of Threat," in *The Affect Theory Reader*, Melissa Gregg and Gregory J. Seigworth, eds. (Durham, NC: Duke University Press): 52–70.

Mattingly, C., Lawlor, M., and Jacobs-Huey, L. (2002), "Narrating September 11: Race, Gender, and the Play of Cultural Identities," *American Anthropologist*, 104, 3: 743–753.

Mbembe, Achille (2002), "African Modes of Self-Writing," trans. Steven Rendall, *Public Culture*, 14, 1: 239–273.

Mearsheimer, John (1990), "Back to the Future: Instability in Europe after the Cold War," *International Security*, 15, 1: 1–56.

Mearsheimer, John (1994–5), "The False Promise of International Institutions," *International Security*, 19, 3: 5–49.

Mearsheimer, John (2001), *The Tragedy of Great Power Politics* (New York: Norton).

Mearsheimer, John (2002), "Conversation with John Mearsheimer," Institute of International Studies, UC Berkeley, Harry Kreisler interviewer, April 8: http://globetrotter.berkeley.edu/people2/Mearsheimer/mearsheimer – con1.html. Accessed September 12, 2011.

Mearsheimer, John and Stephen Walt (2006), "The Israel Lobby and American Foreign Policy," *Middle East Policy*, 13, 3: 29–87.

Mearsheimer, John and Stephen Walt (2007), *The Israel Lobby and American Foreign Policy* (New York: Farrar, Straus and Giroux).

Mearsheimer, John and Stephen Walt (2009), "Is It Love or the Lobby? Explaining American's Special Relationship with Israel," *Security Studies*, 18, 1: 58–78.

Melander, E., M. Oberg, and J. Hall (2007), "The 'New Wars' Debate Revisted: An Empirical Evaluation of the Atrociousness of 'New Wars,'" *Uppsala Peace Research Papers*, 9: 1–42.

Mercer, Jonathan (2005), "Rationality and Psychology in International Politics," *International Organization*, 59, 1: 77–106.

Mercer, Jonathan (2010), "Emotional Beliefs," *International Organization*, 64: 1–31.

Mertus, Julie (2000), *War's Offensive on Women: The Humanitarian Challenge in Bosnia, Kosovo, and Afghanistan* (San Francisco: Kumarian Press).

Millennium: Journal of International Studies (2008), "Roundtable Discussion: Reflections on the Past, Prospects for the Future in Gender and International Relations," with Marysia Zalewski, Ann Tickner, Christine Sylvester, Margot Light, Vivienne Jabri, Kimberly Hutchings and Fred Halliday, 37, 1: 153–179.

Mohanty, Chandra (2003), *Feminism Without Borders: Decolonizing Theory, Practicing Solidarity* (Durham, NC: Duke University Press).

Mohanty, Chandra, Minnie Bruce Pratt and Robin Riley, eds. (2008), *Feminism and War: Confronting US Imperialism* (London: Zed Press).

Mookherjee, Nayanika (2011), "'Never Again': Aesthetics of 'Genocidal' Cosmopolitanism and the Bangladesh Liberation War Museum," *Journal of the Royal Anthrpological Institute*, 1, issue supplement S1(May): S71-S91.

Moon, Katherine (1997), *Sex Among Allies: Military Prostitution in US–Korea Relations* (New York: Columbia University Press).

Morgenthau, Hans (1965), *Politics Among Nations: The Struggle for Power and Peace* (New York: Alfred Knopf).

Moser, Caroline and Fiona Clark, eds. (2001), *Victims, Perpetrators, or Actors? Gender, Armed Conflict, and Political Violence* (London: Zed Books).

Mouffe, Chantal (2002), *Politics and Passion: The Stakes of Democracy* (London: Centre for the Study of Democracy).

Mueller, J. (2004), "Why Isn't There More Violence?" *Security Studies*, 13, 3: 191–203.

Mundy, Jacob (2011), "Deconstructing Civil Wars: Beyond the New Wars Debate," *Security Dialogue*, 42, 3: 279–295.

Nafisi, Azar (2004), *Reading Lolita in Tehran* (London: Fourth Estate).

Nakamura, Jeanne and M. Mihaly Csikszentmihalyi (2002), "The Concept of Flow," in C. R. Snyder and S. J. Lopez, eds. *Handbook of Positive Psychology* (Oxford: Oxford University Press): 89–105.

Ndibe, Okey and Chenjerai Hove, eds. (2009), *Writers, Writing on Conflicts and Wars in Africa* (London: Adonis & Abbey Publishers).

Neumann, Iver (2008), "The Body of the Diplomat," *European Journal of International Relations*, 14, 4: 671–695.

Newman, Edward (2004), "The 'New Wars' Debate: A Historical Perspective is Needed," *Security Dialogue*, 35, 2: 173–189.

Newman, Edward (2009), "Conflict Research and the 'Decline' of Civil Wars," *Civil Wars*, 11, 3: 255–278.

Nordstrom, Carolyn (1995), "Creativity and Chaos: War on the Frontlines," in *Fieldwork Under Fire: Contemporary Studies of Violence and Survival*, Carolyn Nordstrom and Antonius Robben, eds. (Berkeley, CA: University of California Press): 129–154.

Nordstrom, Carolyn (1997), *A Different Kind of War Story* (Philadelphia, PA: University of Pennsylvania Press).

Nordstrom, Carolyn (2004), *Shadows of War: Violence, Power, and International Profiteering in the Twenty-First Century* (Berkeley, CA: University of California Press).

Nussbaum, Martha (1990), *Love's Knowledge: Essays on Philosophy and Literature* (Oxford: Oxford University Press).

Nussbaum, Martha (1999), "Professor of Parody," *The New Republic*, 22, 2: 37–45.

Nussbaum, Martha (2001), *Upheavals of Thought: The Intelligence of Emotions* (Cambridge: Cambridge University Press).

Nussbaum, Martha (2003), *Thinking about Feeling: The Intelligence of Emotions* (Cambridge: Cambridge University Press).

Oakley, Ann (1998), "Gender, Methodology and People's Ways of Knowing: Some Problems with Feminism and the Paradigm Debate in Social Science," *Sociology* 32, 4: 707–731.

Okin, Susan Moller, (1999), "Is Multiculturalism Bad For Women?", in Joshua Cohen, Matthew Howard and Martha Nussbaum, eds., *Is Multiculturalism Bad for Women?* (Princeton, NJ: Princeton University Press): 7–26.

Orford, Anne (2003), *Reading Humanitarian Intervention: Human Rights and the Use of Force in International Law* (Cambridge: Cambridge University Press).

Orford, Anne (2011), *International Authority and the Responsibility to Protect* (Cambridge: Cambridge University Press).

Parashar, Swati (2009), "Feminist International Relations and Militant Women: Case Studies from Sri Lanka and Kashmir," *Cambridge Review of International Affairs*, 22, 2: 235–256.

Parashar, Swati (2011), "Embodied 'Otherness' and Negotiations of Difference," in Christine Sylvester, ed., "Forum: Emotion and the Feminist IR Researcher," *International Studies Review*, 13, 4: 696–699.

Park-Kang, Sungju (2011), "Gendered Agency in Contested Truths: The Case of Hyunhee Kim (KAL858)", in Linda Åhäll and Laura J. Shepherd (eds.), *Gender, Agency and Political Violence* (London: Palgrave: 115–131).

Parpart, Jane and Marysia Zalewski, eds. (2008), *Rethinking the Man Question: Sex, Gender, and Violence in International Relations* (London: Zed).

Patel, Rajeev and Philip McMichael (2004), "Third Worldism and the Lineages of Global Fascism: The Regrouping of the Global South in the Neoliberal Era," *Third World Quarterly*, 25: 231–254.

Pateman, Carole (1988), *The Sexual Contract* (Stanford, CA: Stanford University Press).

Peck, James (2010), *How the US Government Co-opted Human Rights* (New York: Metropolitan Books).

Penttinen, Elina (2007), *Globalization, Prostitution, and Sex Trafficking: Corporeal Politics* (London: Routledge).

Peterson, V. Spike, ed. (1992), *Gendered States: Feminist (Re)Visions of International Relations Theory* (Boulder, CO: Lynne Rienner Press).

Posen, Barry (1993), "The Security Dilemma and Ethnic Conflict," *Survival*, 35, 1: 27–47.

Pouliot, Vincent (2008), "The Logic of Practicality: A Theory of Practice of Security Communities," *International Organization*, 62 (Spring): 257–288.

Raghuram, Parvati, Clare Madge, and Pat Noxolo (2009), "Rethinking Responsibility and Care for a Postcolonial World," Geoforum, 40, 1: 5–13.

Rose, Gillian (1996), *Mourning Becomes the Law* (Cambridge: Cambridge University Press).

Rowe, Cami (2013), *Anti-War Actions and the Performative Construction of the War on Terror* (London: Routledge).

Saeidi, Shirin and Heather Turcotte (2011), "Politicizing Emotions: Historicizing Affective Exchange and Feminist Gatherings," in Christine Sylvester, ed., "Forum: Emotion and the Feminist IR Researcher," *International Studies Review*, 13, 4: 693–695.

Sandel, Michael (1998) *Liberalism and the Limits of Justice*, 2nd edition (Cambridge: Cambridge University Press).

Scarry, Elaine (1985), *The Body in Pain: The Making and Unmaking of the World* (Oxford: Oxford University Press).

Schott, Robin May (2011), "Pain, Abjection, and Political Emotion," *Passions in Context; International Journal for the History and Theory of Emotions*, no. 2: 7–34.

Scott, Joan (1991), "The Evidence of Experience," *Critical Inquiry*, 17, 4: 773–797.

Senfft, Alexandra (2008), *Schweigen tut Weh: Eine deutsche Familiengeschichte* (Ullstein Taschenbuchvlg).

Shane, Scott, Mark Mazzetti and Robert F. Worth (2010), "A Secret Assault on Terror Widens on Two Continents," *New York Times*, August 15, pp. 1, 10–11.

Shapiro, Michael (2011), "The Presence of War: Here and Elsewhere," *International Political Sociology*, 5, 2: 109–125.

Singer, P.W. (2006), *Children at War* (Berkeley, CA: University of California Press).

Singer, P.W. (2009), *Wired for War: The Robotics Revolution and Conflict in the 21st Century* (New York: Penguin Books).

Sjoberg, Laura, ed. (2010), "Introduction," *Gender and International Security: Feminist Perspectives* (New York: Routledge).

Sjoberg, Laura (2011), "Gender, the State, and War Redux: Feminist International Relations across the 'Levels of Analysis.'" *International Relations* 25, 1: 108–134.

Sjoberg, Laura and Caron Gentry (2007), *Mothers, Monsters, Whores: Women's Violence in Global Politics* (London: Zed).

Skjelsbæk, Inger (2012), *The Political Psychology of War Rape: Studies From Bosnia and Herzegovina* (London: Routledge).

Slater, Jerome (2009), "The Two Books of Mearsheimer and Walt," *Security Studies*, 18, 1: 4–57.

Smith, Steve (2004), "Singing Our World Into Existence: International Relations Theory and September 11," *International Studies Quarterly*, 48, 3: 499–515.

Snyder, Richard, H.W. Bruck, Burton Sapin and Valerie Hudson (2002), *Foreign Policy Decision-Making (Revisited)* (London: Palgrave).

Spelman, Elizabeth (1997), *Fruits of Sorrow: Framing Our Attention to Suffering* (Boston: Beacon Press).

Spivak, Gayatri Chakravorty (1988), "Can the Subaltern Speak?," in Cary Nelson and Lawrence Grossberg, eds., *Marxism and the Interpretation of Culture* (Chicago: University of Illinois Press): 271–313.

Stanger, Allison (2009), *One Nation Under Contract: The Outsourcing of American Power and the Future of Foreign Policy* (New Haven: Yale University Press).

Stephens, Julie (1989), "Feminist Fictions: A Critique of the Category 'Non-Western Woman' in Feminist Writings on India," in Ranajit Guha, ed., *Subaltern Studies, 6: Writings on South Asian History and Society* (Delhi: Oxford University Press): 92–126.

Stern, Maria (2005), *Naming Security – Constructing Identity: 'Mayan Women' in Guatemala on the Eve of 'Peace'* (Manchester: Manchester University Press).

Stern, Maria (2006), "'We' the Subject: The Power and Failure of (In)Security," *Security Dialogue*, 37, 2: 187–205.

Stiehm, Judith (1984), *Women's and Men's Wars* (Oxford: Pergamon Press)

Svensson, Isak and Mathilda Lindgren (2011), "From Bombs to Banners: The Decline of Wars and the Rise of Unarmed Uprisings in East Asia," *Security Dialogue*, 42, 3: 219–237.

Sylvester, Christine (1994a), *Feminist Theory and International Relations in a Postmodern Era* (Cambridge: Cambridge University Press).

Sylvester, Christine (1994b), "Empathetic Cooperation: A Feminist IR Methodology," *Millennium: Journal of International Studies*, 23, 2: 315–336.

Sylvester, Christine (2000), *Producing Women and Progress in Zimbabwe: Narratives of Identity and Work from the 1980s* (Portsmouth, NH: Heinemann).

Sylvester, Christine (2004), "Gendered Development Imaginaries: Shall We Dance, Pygmalion?" in *Feminist International Relations: An Unfinished Journey* (Cambridge: Cambridge University Press): 224–241.

Sylvester, Christine (2005), "The Art of War/The War Question in (Feminist) IR," *Millennium*, 33, 3: 855–878.

Sylvester, Christine (2006), "Bare Life as a Development/Postcolonial Problematic," *Geographical Journal*, 172, 1–2: 66–77.

Sylvester, Christine (2007), "Whither the International at the End of IR?," *Millennium: Journal of International Studies*, 35, 3: 551–573.

Sylvester, Christine (2009), *Art/Museums: International Relations Where We Least Expect It* (Boulder: Paradigm Publishers).

Sylvester, Christine (2010a), "Tensions in Feminist Security Studies," *Security Dialogue*, 41, 6: 607–614.

Sylvester, Christine (2010b), "War, Sense, and Security," in Laura Sjoberg, ed., *Security Studies: Feminist Contributions* (New York: Routledge): 24–37.

Sylvester, Christine, ed. (2011a), *Experiencing War* (London: Routledge).

Sylvester, Christine, ed. (2011b), "Forum: Emotion and the Feminist IR Researcher," *International Studies Review*, 40, 4: 687–708.

Sylvester, Christine (2011c), "Writing Emotion," in "Forum: Emotion and the Feminist IR Researcher," *International Studies Review*, 40, 4: 703–707.

Sylvester, Christine (2011d), "Development and Postcolonial Takes on Biopolitics and Economy," in Jane Pollard, Cheryl McEwan, and Alex Hughes, eds., *Postcolonial Economies: Rethinking Material Lives* (London: Zed): 185–204.

Sylvester, Christine and Swati Parashar (2009), "The Contemporary 'Mahabharata' and the Many 'Draupadis': Bringing Gender to Critical Terrorism Studies," in Richard Jackson, Marie Breen Smyth, and Jeroen Gunning, eds., *Critical Terrorism Studies: A New Research Agenda* (London: Routledge): 178–193.

Taylor, Brandon (2004), *Collage: The Making of Modern Art* (London: Thames and Hudson).

Tickner, Ann (1988), "Hans Morgenthau's Principles of Political Realism: A Feminist Reformulation," *Millennium: Journal of International Studies*, 17, 3: 429–440.

Tickner, Ann (1992), *Gender in International Relations: Feminist Perspectives on Achieving Global Security* (New York: Columbia University Press).

Tickner, Ann (1997), "You Just Don't Understand: Troubled Engagements Between Feminists and IR Theorists," *International Studies Quarterly*, 41, 4: 611–632.

Tolstoy, Leo (1957) [1865–68]) *War and Peace* (London: Penguin).

Turkle, Sherry, ed. (2007), *Evocative Objects: Things We Think With* (Cambridge, MA: MIT Press).

Utas, Mats (2005), "Victimcy, Girlfriending, Soldiering: Tactic Agency in a Young Woman's Social Navigation of the Liberian War Zone," *Anthropological Journal*, 78, 2: 403–430.

Van Evera, Stephen (1994), "Hypotheses on Nationalism and War," *International Security*, 9, 1: 58–107.

Vasquez, John (2009), *The War Puzzle Revisited* (Cambridge: Cambridge University Press).

Volger, Candace (2004), "Much of Madness and More of Sin: Compassion, for Ligeia," in Lauren Berlant, ed., *Compassion: The Culture and Politics of an Emotion* (New York: Routledge): 29–58.

Waller, Gregory (2010), *The Living and the Undead: Slaying Vampires, Exterminating Zombies* (Champaign, IL: University of Illinois).

Walt, Stephen (1999), "Rigor or Rigor Mortis? Rational Choice and Security Studies," *International Security*, 23, 4: 5–48.

Walt, Stephen (2009), "Why They Hate Us (II): How Many Muslims Has the US Killed in the Past 30 Years?" *Foreign Policy*, November 30: foreignpolicy.com.

Walt, Stephen (2011), "Is America Addicted to War?" *Foreign Policy*, April 4, accessed online May 2: foreignpolicy.com.

Weldes, Jutta (1999), *Constructing National Interests: The United States and the Cuban Missile Crisis* (Minneapolis, MN: University of Minnesota Press).

Wibben, Annick (2010), *Feminist Security Studies: A Narrative Approach* (London: Routledge).

Wilson, Margaret, ed. (1969), *The Essential Descartes* (New York: Mentor).

Winnubst, Shannon (2003), "Vampires, Anxieties and Dreams: Race and Sex in the Contemporary United States," *Hypatia*, 18, 3: 1–20.

Wolfers, Arnold (1962), *Discord and Collaboration: Essays on International Politics* (Baltimore, OH: Johns Hopkins University Press).

Wood, Elizabeth Jean (2011), "Rape Is Not Inevitable During War," in Kathleen Kuehnast, Chantal de Jonge Oudraat, and Helga Hernes, eds., *Women and War: Power and Protection in the 21st Century* (Washington: The United States Institute of Peace Press): 37–63.

Wright, Quincy (1965), *A Study of War. 2nd Edition with a commentary on war since 1942* (Chicago: University of Chicago Press).

Zarkov, Dubravka (2007), *The Body of War: Media, Ethnicity, and Gender in the Break-Up of Yugoslavia* (Durham, NC: Duke University Press).

Zenko, Micah (2010), *Between Threats and War: US Discreet Military Operations in the Post-Cold War World* (Stanford, CA: Stanford University Press).

INDEX

9/11: and grief reactions 102, 106; reactions by African-American mothers 57; role of IR in 112

abducted women, returning home 71, 72
abstraction, war as 1–2, 17
Abu Ghraib 70, 81
Actors: expanding those considered actors in war 9; 'wrong' actors in fall of Berlin Wall 18
Adichie, Chimamanda Ngozi (author *Half of a Yellow Sun*) 101–2, 106–8
affect and affect theory 6, 93, 95–9, 102, 108–10, 123
Afghanistan: and bodily injury 67; and compassion 102; endless nationalist war 42; involvement of US offers little in gains 19; recycling weapons 4; war moving across social realms 2
Africa: Congo 2–3, 24–5, 82–4; Liberia 2–3, 23–4; Sierra Leone 2–3, 54–7, 72; Zimbabwe 24, 42, 51, 129n
aftermath of war 25, 28
Agamben, Giorgio: bare life contexts 69, 70, 72, 107; biopolitical analysis 69–70; exceptional violence 27, 72, 124
agency: in constructivism 26; human body as agent and target 65–86; Sri Lankan women 54; women's agency not recognized in Sierra Leone 56
Ahmed, Sara, on emotions 88, 89, 93–4

Alison, Miranda: interview-based study of female combatants 52–4; use of narratives 57
Allison, Graham, on the Cuban missile crisis 29
Al Qaeda 2, 4, 30
anti-war position, and feminism 60; *see also* peace
Arab Spring 2, 9, 123
Arendt, Hannah: emotions as anti-political 129n; political sociology 35
art and literature: blurring the line between fact and fiction 12, 120–1, 125; depicting descent into the ordinary 71–2; depicting emotions in war 100; and invented persons 12; and pain 117; used as sources for war as experience research 40, 118–23, 125
Art/ Museums: International Relations Where We Least Expect It (Sylvester, 2009) 126
A Sunday at the Pool in Kigali (Courtemanche, 2003) 71–2, 107
A Woman in Berlin (anonymous) 119, 129n

Baaz, Maria Eriksson: on post-war rape counselling 103; on rape during Congo war 24–5, 82–4
Balkan wars 2, 80–2, 121, 129n
Bananas, Beaches and Bases: making Feminist Sense of International Relations (Enloe, 1989) 42–5
Bangladesh 123–4

bare life contexts 69, 70, 72, 107
Barkawi, Tarak: and the need to study social relations of war 25; postcolonial critical IR approach 29–32, 60
Bennett, Jane, on nonhuman actants 5, 76
Berlant, Lauren: on compassion 100–2, 104; on definitions of 'experience' 48, 95; on social sciences' aversion to people 111
Berlin Wall 9, 18
Biafra War 101–2, 107–8
bin Laden, Osama 2, 17, 32
biopolitics 69–70
Blanchard, Eric, and the English School 128n
Bodies; bodily senses 78–9; the body and experience 49; body-mind linking 6, 65, 87, 97, 109; as contested entity 5, 73–9; as cyborgs 75–6; emotions not located in 93–5, 109; feminist case studies of bodily injury 79–86; and mourning 105–6; as the object of study 5; physical suffering in war 101; as social constructions 78; starting with the body as way to new analyses 62; war as physical experience 65–86; women's bodies vs men's and fitness to fight 38; zombies 76–7; *see also* rape
Bordo, Susan: on emotions around women's bodies 87; on sexed bodies 74
Bosnia 80–1
Braidotti, Rosie, on sexed bodies 74
Brighton, Shane: on philosphical-ethical abstract framework for experience of war 85–6, 116; on security studies 10; on war as form of social relations 31–2
Brown, Wendy, on the 'state' 115
Burundi 12, 68–9
Bush, George W.: on grief 106; influencing the feelings of citizens 96; misrepresenting the central content of war 67
Butler, Judith: on 'affect-as-ideology' 109; on bodies being constructed socially 78, 109; on 'collateral damage' 1; on emotion propelling politics 105; impact on feminist IR war studies 50, 61; on mourning 40, 78, 102–4, 105–6, 123, 126; on 'sex' and 'gender' 73, 77–8; on suffering 104; on war as a social institution 109; on war creating communities 118; on whose stories count? 45

causes of war 21–2
Cebrowski, Admiral 113
Chan, Stephen, critique of 'new war' thinking and irrationality 24–5
children, involvment in wars generally 20, 55
child soldiers: children claiming identity as 84; females treated differently to males post-conflict 56; in new wars 2–3, 23; study methods 118
Code Pink (Rowe, 2013) 128n
cold war 9, 18, 22
collages of war 3, 125–6
'collateral damage: in computerised warring 67; as hideous concept 124; people's experiences as 17
combatants, female 52–7, 75
'combat flow' 106
compassion 100–2, 104–5
Congo: child soldiers 2–3; rape 3, 24–5, 82–4
conjugal order (disrupting) 55–6
Connolly, William: and affect studies 6, 97–9, 109; relays 6, 88, 97–9, 116
conscientious objectors 41
constructivist IR thinking 10, 26–8, 33
Courtemanche, Gil (author *Sunday at the Pool in Kigali*) 71–2
Crawford, Neta, on emotions 89–91, 95
criminal networks involved in wars 23
critical IR analysis 10, 28–34, 43, 69; *see also* feminist IR takes on war
Cuban missile crisis 29
cyborgs 5, 124; bodies as 75–6

Damasio, Anthony, on affect 96
Das, Veena: on communicating pain 117; on ordinary vs extraordinary 14, 37, 70–1, 72–3, 88
D'Costa, Bina, on research *in situ* 51
democratic model, idealized by west 30
Deogratias (character in study) 68–9, 73
Der Derian, James: critical IR analysis 32–5; criticism of constructivist IR thinking 33; crossover work 125; emotions of researchers 128n; and human behavior at heart of warfare 113; imagery 33–4; male-dominated theories of war 66; methodologies 53; war studies 60
'descent into the ordinary' 70–1
development studies: critical development studies 104–5; de-development/ underdevelopment as strategy 23–4; development experts as spectators 105

Dillon, Mick, and biopolitical analysis 69
discourse analysis 52, 57–9
domestic tasks, enabling the work of elite
  men 42; *see also* home front, women
  working on
DRC (Democratic Republic of Congo):
  child soldiers 2–3; rape 3, 24–5, 82–4
Drezner, Daniel, on zombies 76
drones and robots 2, 32
Duffield, Mark, on the search for causes of
  war 21
Dutch government 45–9

Elshtain, Jean: feminist perspective on
  war 39–42; on how bodies of sexed
  women are meant to be outside
  military engagement 74; on women's
  empowerment by war 82
emotion; definition of 13, 90–1; emotions
  of researchers 53, 91–2; new wars carved
  out of manufactured animosity 22–5;
  positive emotions and war 106–8; war as
  emotional experience 5–6, 87–110
empowerment, war as 56
Enloe, Cynthia: anti-war position 44–5;
  feminist IR takes on war 42–5; on
  gendered emotion 109; on Iraq 44, 45;
  on khaki 7, 39; on militarization 42–5,
  74, 98; reason for naming her book
  The Morning After... 25
enthusiasm for war 106–8
ethics, research 51
ethnicity, gendered 80–2, 84
Europe: and new war thinking 24–5; wars
  elsewhere seen as 'small wars' 30
experiences of war: compared to statistics
  on war 111; and discourses 52; and
  emotions 95–110; injury, war is about
  3–4, 66–73, 114–15; studying women's
  war experiences 49–52; whose and what
  to study 45–9
Experiencing War workshops 31–2, 36,
  127n

false consciousness 56, 106
Female Soldiers in Sierra Leone
  (MacKenzie, 2013) 54–7
feminist IR takes on war 6–10, 38–62,
  112–13, 114
Feminist Security Studies: A Narrative
  Approach (Wibben, 2010) 57
feminist theory, basing research in 59
Ferguson, Kathleen, on women's
  experiences of war 48

Fiction: blurring the line between fact
  and fiction 12, 120–1, 125; depicting
  descent into the ordinary 71–2; depicting
  emotions in war 100; used as sources for
  war as experience research 40, 118–23, 125
fieldwork *see* methodologies
Fierke, Karin: constructivist ideas of social
  aspects of war 27–8; on emotion 88, 94
Finnemore, Martha, and constructivism
  26–7
flow (being the the 'zone') 106–7
Foucault, Michel 4, 29, 69, 70, 71, 124
Fox Girl (Keller, 2002) 119–20, 122
Frames of War (Butler, 2010) 22–5
Friedan, Betty, on emotions 88
Fukuyama, Francis: on the decade between
  cold war and war on terror 43; on the
  end of history 22

gangs 23, 69, 115
gender: gender relations in IR 38; *see also*
  feminist IR takes on war; sex vs gender
  73
Germany, constructivist views of WWI and
  WWII 27–8
Gibbon, Jill, artwork of people at arms fairs
  118
Gilda Stories, The (Gomez, 1991) 85
'Girls Left Behind' (UNICEF program)
  56, 84
Globalization: and arms trading 43; critical
  IR analysis 43; and diffusion of ideas
  23; and increased political violence
  22–3; increases ways that people access
  war 4–5; and new wars 22; socially
  connective of enemies and friends 43;
  and trade 23; wars as integral aspects of
  31
Global South 29; *see also* postcolonialism
Goldstein, Joshua: on increase of peace in
  the world 111; on war and sexed bodies
  of men 74–5
Grant, Judith, on experience as measure of
  oppressive situations 47
Greeks (Ancient) and war 40
grey zones of war 117–24
grief *see* mourning
Gronemeyer, Marianne, on compassion
  104–5
Grossberg, Lawrence, on affect 108–10
Grosz, Elizabeth, on sexed bodies 74
Guantanamo camp 70
Guatemala 52
Guernica (Picasso) 126, 130n

Half of a Yellow Sun (Adichie, 2007) 101–2, 106–8
Hallinan, Joseph, on group errors 37
Hansen, Lene: on security studies 10; on Yugoslavia 121, 129n
Harari, Yuval Noah, on war satisfaction 106
Haraway, Donna, on cyborgs 75–7
Harding, Sandra, on feminist mediations 48
heroism, female 42
Hirsi Ali, Aayan (Somali-Dutch feminist) 7, 8, 45–9, 88
Holzner, Brigitte, on women in the Liberian war 116
home front, women working on 40–1
home spaces, war crossing 35–6, 61, 72
homosexuality in the military 38, 74
households and experience of war 35–6
Hove, Chenjerai, on fact/fiction 12
humanitarian interventions 27
humanitarianism 115
humanitarian law 21, 23
human rights 4, 21, 23, 24
Huntington, Samuel, on the West and the Rest perspective 24
Hussein, Saddam 2, 30, 32, 33, 96, 119
Hutchings, Kimberley, on the moral value of war 84

Identity: and experience 49; multiple identities 47; researchers adopting multiple 92–3; women's identities forced by communities 51
India 3
injury, war is about 3–4, 66–73, 113–15; *see also* rape
interdisciplinary approaches to war studies 31–2
international institutions (IOs) 21
international relations (IR): differences between IR and feminist IR 61; dwarfed by security studies in recent years 60; international components of all modern wars 4; not good at studying people in war 111–12; operating at an abstract level 1–2, 17; position in academia 67; and pro/anti-war stances 8–9; takes on war 17–37; as a war participant 112
intersectional analyses 9–10, 47–8, 124–5
interviews (research method) 50–2, 55, 57
IRA 54
Iran 8, 119
Iraq: and bodily injury 67; and compassion 102; involvement of US offers little in gains 19; lessons from women (Enloe)

44–5; media-entertainment networks and US decision to invade 32; state-on-state 4; war moving across social realms 2
Islam: and Bosnian women 81; burdens on women (Hirsi Ali) 46
Israel: collective violence 2; Israeli-Palestinian relations described in terms of people, not states 20; Israel lobby 19–21; and US 19–21

Jabri, Vivienne: on attention to classical texts 129n; on war as chief international political matrix of our time 5, 43–4
joy, in war 106–8

Kaldor, Mary: on globalization 31; on new war thinking in Europe 22–5
Kashmir 92, 114
Keller, Nora Okja (author of *Fox Girl*) 119–20, 122
khaki 7, 39
Kidder, Tracy, on bodily injury 68–9
kidnappings 24
killing: 'collateral damage' 17, 67, 124; deaths are seen as 'self-evident' and not the content of war 68; states making exceptions to their own laws 69–70; women carrying out 47, 56
Korea 3, 58, 60, 119–20, 124
Kronsell, Annica, on interviews as research method 50
Kwon, Heonik: research methods 118, 124; on social relations in cold war in Vietnam 25, 28; use of fiction to research war 118

language, use of 29, 117
Lebanon war 2006 20
Levinas, Emmanuel, on how war remakes people's identities 31
Leys, Ruth, on affect studies 95–6, 97
Liberia: blood diamonds 23–4; child soldiers 2–3
Libya 2, 4, 21
Lipschutz, Ronnie, on the 'state' 115
literature *see* art and literature
Lolita (Nabokov, 1955) 8, 121
looting 123–4
LTTE 53–4

MacGinty, Roger, on museum looting 123
MacKenzie, Megan: on biopolitical analysis 70; on exceptionality of women's participation 75; feminist IR take on

war 54–7; researchers' emotions 92; on women's war duties 72

male-dominated theories of war 38, 41, 74

Malesevic, Sinisa, and contemporary sociology 35

Manning, Erin: on bodily senses 78–9, 84–5, 105; crossover work 125

maps and models, use of 68

Marshall, Sandra, and emotions of war researchers 91–2

Massumi, Brian, on affect 96–7

McMichael, Philip, on 'the state' 128n

Mearsheimer, John: on emotions 94; on Iraq war 19; personal experience of war 67; realist IR war thinking in US 18–22, 113; and security studies 60; on state powers 19

media: and international relations 126; media-entertainment networks and US decision to invade Iraq 32; people experiencing war through media 5, 98; social networking media and invented persons 12; as some people's main experience of war 98

mediation, of knowledge 48

Methode (character in novel) 72, 107, 124

Methodologies: emotions of researchers 53, 91–2; ethnographic methods 47, 53, 59, 91; exemplary text methodology 3; experiences as a methodology 48; gender biases in scientific approaches 61; interviews 50–2, 55, 57; narrative/ discourse analysis 52, 57–9; non-interview based ways to research experiences 52, 57–9, 118; quantitative analysis 51; quasi-anthropological 32; sampling 53; social science methodologies 29; studying up 2, 13–14, 47, 86, 99, 109–11, 115, 125

micropolitics 98, 129n

Middle East: Arab Spring 2, 9, 123; Iran 8, 119; Lebanon war 2006 20; see also Iraq; Israel

militarization 7, 42–5, 60, 98

military organizations: and the presence of women 38, 74, 83; and women being 'allowed' to join by men 75

MIME-NET 32, 34, 35

misrepresentation of central content of war 67–8

Mookherjee, Nayanika, on museum objects 123

Moon, Katherine: on military 'entitlement' to local girls 120; and narrative analysis of women's experiences 58

The Morning After: Sexual Politics at the End of the Cold War (Enloe, 1993) 25

Motherhood: feminist maternal thinking 42; good mothers linked to good soldiers 40, 41; symbolic capacities of the maternal body 80; and women going off to fight 38

Mouffe, Chantal, on rationality 115

Mourning: mourning as politics 102–4; need for 28; potential for transnational politics 50; women as designated mourners 40

Mugabe, Robert, and blood diamonds 24

Mullen, Admiral Michael 118–19

museum zones of war 123–4

Nafisi, Azar (author of Reading Lolita in Tehran) 8, 119, 121

narrative analysis 52, 57–9

nationalism: no longer the basis of wars 43; as a political phenomenon 42–3; and Sri Lankan women 53–4

Nazis 27, 28

Neuropolitics (Connolly, 2002) 97–8

neuroscience 95–6

new wars: based around political ideologies 23–4; and bodily injury 67; carved out of manufactured animosity 22–5; current conflicts 2–3; Mary Kaldor on new war thinking in Europe 22–5; new war thinking and non-state military functions 115; new war thinking in Europe 22–5; not always featuring rape 82; ordinary civilians as spectators and sufferers 103; targetting civilians 128n; and women and children 39, 55, 82; and women's bodies 82

Nimo's War, Emma's War (Enloe, 2010) 52

nonconscious identity 96

Nordstrom, Carolyn: on grey zones 122–3; on interviews as research method 50–1; narrative analysis of women's experiences 58; on the need for feminist mediated knowledge 48

Northern Ireland, female combatants 54

novels about war see fiction

Nussbaum, Martha: on bodies being constructed socially 78; on emotions 95; on fictional writing vs philosophy 120–1

Obama, Barack 68, 115
objectivity in social sciences 92
Okin, Susan, on beseiged cultural groups 46
ordinary people: 'descent into the ordinary'
  70–3; emotions of ordinary people not
  considered 91; fiction depicting descent
  into the ordinary 71–2; Israeli-Palestinian
  relations described in terms of people,
  not states 20; ordinary civilians as
  spectators and sufferers in new wars 103;
  in war as a social institution 4–5
Orford, Ann: and *Guernica* 129n; and
  Responsibility to Protect 21

pain: creating community 118; and
  language 117
Pakistan 2, 4, 42, 123
Palestine: collective violence 2; endless
  nationalist war 42; and the Israel lobby
  20
Parashar, Swati: and location of emotions
  94–5; on researchers' emotions 92–3
Park-Kang, Sungju, and critical IR thinking
  29
partition (South Asia) 70–1
Passion of World Politics (Crawford, 2000)
  89
Patel, Rajeev, on 'the state' 128n
Peace: constructivism links with peace
  resarch 27; decline of war in the world
  1, 111; new feminist thinking doesn't
  emphasise 50, 60; peace dividends/
  democratic peace 22; 'peacegaming' 33;
  peace institutions 21; realm of peace
  conjoined with realm of war 41; as usual
  feminist object of study 6–7, 8, 10, 60;
  wars do not always end when IR says
  they do 25, 28; women and peace 39
Peck, James, on human rights 127n
Penttinen, Elina, and researchers' emotions
  129n
'personal is political' 47, 91, 93
phallic power 74
philosophies of war 40
political sociology 9, 10
Politics of Touch: Sense, Movement,
  Sovreignty (Manning, 2007) 78
positivism 10
postcolonialism and development issues
  23–4; and emotions 94; and globalization
  43; as part of critical IR 10; *see also*
  Barkawi, Tarak
post-conflict processes 56, 84
postmodernist constructivism 26

poststructuralism 10, 45, 47, 48, 58 ;
  location of 70; manifestation in gender
  stories 48; nonstate forms of 70–1; and
  objectivity 29; *see also* Connolly, William;
  Der Derian, James
prevention of war 21
private militaries/ security forces 2, 23, 69
proximities to war 101–2, 110
public/ private spheres, gender and war
  politics 35–6

race 47; *see also* ethnicity, gendered
Ranciere, Jacques, on politics not being
  spatial 35–6
Rape: in DRC 3, 4, 24–5, 43, 82–4; and
  'just' warriors 41; of men 81; in new wars
  23, 67; as physical experience of war
  79–83; raped women in Congo
  want money for children rather than
  counselling 103; and rationality 24–5
rationality: in the critical IR tradition 28–9;
  disassociated from body and emotion
  65–6; and emotions working with it
  99; irrationality of 'new wars' 22–5; and
  micropolitics 129n; pervasive assumption
  of 115–16; and rape 24–5; and realism
  19–21; and researchers' emotions 92;
  trying to understand rationalities 24–5;
  vs emotions 87, 89; as a Western ideal
  29–30
Reading Lolita in Tehran (Nafisi, 2004) 8, 119,
  121
realist approaches to war: compared to
  constructivism 26; and the injury of
  people 113; and militarization 43; and
  rationality 19–20; realist IR war thinking
  in US 18–22; and the 'state' 10
Reid, Julian, and biopolitical analysis 69
relays 6, 88, 97–9, 116
religion: and feminism 45–6; and gendered
  ethnicity 81–2; as part of identity
  repertoire 47
researchers' emotions 53, 91–2
research methods *see* methodologies
respect, women seeking 54
'Responsibility to Protect' 21, 30
robots and drones 2, 32
Rwanda 2, 4, 68–9, 71–2, 109

Saeidi, Shirin, on location of emotions 93,
  94
Sandel, Michael, on liberalism 7
Scarry, Elaine, on bodily injury 4, 66–73,
  114, 117

security studies: feminist security studies
49, 58; and intersectionality 10; shift in
focus towards, criticised 31, 60; taking
precedence over social relations 25
September 11: and grief reactions 102, 106;
reactions by African-American mothers
57; role of IR in 112
Serbia 80–1
sex vs gender 73, 77–8
Shapiro, Michael: crossover work 125;
political sociology 35–6, 61; on space-
body relationships 79, 116
Sheehan, Cindy: anti-war protester 35–6;
and mourning 103
Sierra Leone: child soldiers 2–3; women
and the war 54–7, 72
Sikkink, Katherine, and constructivist IR
thinking 27
Singer, P.W., and sources of evidence 118
'small wars' 29–31
Smith, Steve, on IR as a war participant
112, 113
social anthropology 48, 114
social institution, war as: and 'affect-as-
ideology' 109; affecting culture and
thinking 98; altered social relations
leading to increased political violence
in 21st Century 22–3; cross-nationally
124; and the discipline of international
relations (IR) 17, 33–4, 112; institutional
components 4; and militarization 43–4;
opportunities for women 82; 'retrograde'
social relations in new wars 24; studying
alongside other analyses 9; war as intensified
social connections 50; war as social
activity of collective violence 65–86
social science methodologies 29
sociology, and warfare 35
soldiers, female 52–7, 75
Somalia 2, 7, 23, 45
Soviet Union, post collapse structures 18,
19, 22
space (public vs private domains) 36
spectators, war 101–2, 105, 116
Spelman, Elizabeth, on suffering 6, 102
Sri Lanka: communal violence and women's
experience 48; female combatants 52–4;
militants 2; and nationalism 53–4
standpoint epistemology 47
states: as actors in war, not people 17–22;
centrality of 115; and creation of
international institutions 21; 'new wars'
not based around states 23; no longer the
basis of wars 9; and rationality 128n

stealth wars 2, 23, 32–3
Stern, Maria: narrative analysis of women's
experiences 52, 58; on post-war rape
counselling 103; on rape during Congo
war 24–5, 82–4; on use of naratives 52
Stiehm, Judith, on women as protectees of
war 40
studying up 2, 13–14, 47, 86, 99, 109–11,
115, 125
sufferers, war 101–2
suffragettes 41
A Sunday at the Pool in Kigali
(Courtemanche, 2003) 71–2, 107
Sylvester, Christine: on art/ museums
126; on authorial strategies gesturing
to feminist analysis 77; on biopolitical
analysis 69; on collages 3; on grey zones
of war 117–18; on museum looting
123–4; research methods with women
in Zimbabwe 51, 129n; on studying
up 99; on using fiction 12; on war as
sensory experience for people and
objects 123
Synthetic Theatre of War (STOW) 33

Taliban 2, 23, 30, 42
Taylor, Brandon, on collages 126
Technology: advanced air technology
insulates from reciprocal injury 68;
bodies remain central despite technology
67; 'collateral damage' in computerized
warring 67; drones and robots 2, 32;
military technologies in 'new wars' 23;
surveillance and policing 115; virtuality
in war 32–3
terror, wars against: and rationality 30;
rise of 60; terrorism and 'affective
dusting' 96
touch, importance of 78–9, 85, 105, 123
trauma, as a social experience 27–8
Turcotte, Heather, on location of emotions
93, 94
Turkle, Sherry, on objects 123

UN (United Nations) 21, 30
undead nonhuman-human bodies 5, 76–7
Undoing Gender (Butler, 2004) 77
unnoticed jobs 42
US: constructivist IR thinking 26–7; in
Iraq 19; and Israel 19–21; realist IR war
thinking in US 22–5; response to 9/11
28
Utas, Mats: on child soldiers 84; tactic
agency 85–6, 117

vampires 76
van Gogh, Theo 46
victims, women's victimhood emphasised 81
Vietnam: and bodily injury 67; social relations in cold war in Vietnam 25
violence: collective violence 114–15; exceptional circumstances 72; and modern war theorizing 50; war as social activity of collective violence 65–86; war is about bodily injury 3–4, 66–73, 113–15; women's violence 38–9, 41, 47; *see also* rape
virtuality in war 32–3
Virtuous War: Mapping the Military-Industrial Meda-Entertainment Network (Der Derian, 2009) 32, 53

Walt, Stephen: on emotions 94; on Iraq war 19; realist IR war thinking in US 18–22, 113; on security studies 60; on war as injurious experience 21
war: definition of 3; as valid object of study for feminism 6–7
War and Gender (Goldstein, 2001) 74–5

Weldes, Jutta, on use of language 29
West, Rebecca, on Yugoslavia 121
westernized thinking 29–30
Wibben, Annick, and narrative analysis of women's experiences 57–9
Windsor, (Prince) Harry 128n
A *Woman in Berlin* (anonymous) 119, 129n
Women and Political Violence: Female Combatants in Ethno-National Conflict (Alison, 2009) 52–4
Women and War (Elshtain, 1987) 40–2, 52
Wood, Elizabeth Jean, on rape 83
World War I 27
World War II 27, 42, 110

Yugoslavia 2, 80–2, 121

Zarkov, Dubravka: on gendered ethnicity in the Balkan wars 80–2; narrative analysis 118
Zimbabwe: blood diamonds 24; oppresive new regimes 42; research methods with women 51, 129n
zombies 5, 76–7